T0259747

Lecture Notes in Mathematics

A collection of informal reports and seminars
Edited by A. Dold, Heidelberg and B. Eckmann, Zürich

196

H-spaces

Actes de la réunion de Neuchâtel (Suisse), Août 1970

Publiés par François Sigrist, Université de Neuchâtel, Neuchâtel/Suisse
Textes rédigés en anglais

Springer-Verlag
Berlin · Heidelberg · New York 1971

AMS Subject Classifications (1970): 16 A 24, 55 F 35, 57 F 05, 57 F 25

ISBN 3-540-05461-8 Springer-Verlag Berlin · Heidelberg · New York
ISBN 0-387-05461-8 Springer-Verlag New York · Heidelberg · Berlin

This work is subject to copyright. All rights are reserved, whether the whole or part of the material is concerned, specifically those of translation, reprinting, re-use of illustrations, broadcasting, reproduction by photocopying machine or similar means, and storage in data banks.

Under § 54 of the German Copyright Law where copies are made for other than private use, a fee is payable to the publisher, the amount of the fee to be determined by agreement with the publisher.

© by Springer-Verlag Berlin · Heidelberg 1971. Library of Congress Catalog Card Number 76-159650. Printed in Germany.

Offsetdruck: Julius Beltz, Hemsbach/Bergstr.

Préface

C'est en 1941 que paraissait dans les "Annals of Mathematics" l'article de Heinz Hopf intitulé "Über die Topologie der Gruppen-Mannigfaltigkeiten und ihrer Verallgemeinerungen". Cette date peut être considérée comme celle de la naissance des H-espaces, ou espaces de Hopf. La théorie des H-espaces a imprégné presque toutes les recherches en topologie algébrique depuis son éclosion, et a pris ces dernières années un essor considérable.

Une réunion de spécialistes en topologie algébrique a eu lieu à Chaumont s/Neuchâtel en août 1970, sous les auspices de l'Université de Neuchâtel. Son but était de faire le point des résultats actuels de la théorie, et cette brochure est un compte-rendu des conférences qui y ont été prononcées.

Les congressistes ont eu le plaisir d'accueillir pour une courte visite le professeur Heinz Hopf, ce qui leur a permis de rendre hommage à l'un des pionniers de la topologie moderne. Relevons que bon nombre de ses élèves se trouvaient parmi les participants, preuve supplémentaire de son rayonnement scientifique.

Les organisateurs de la conférence tiennent à remercier ici les institutions qui en ont permis le déroulement harmonieux, en particulier le Fonds national suisse pour la recherche scientifique et l'Etat de Neuchâtel. C'est à ces instances, ainsi qu'à de nombreux autres mécènes, que l'Université de Neuchâtel doit d'avoir pu contribuer de la sorte à la recherche mathématique actuelle.

Neuchâtel, décembre 1970 François Sigrist

C O N T E N T S

FAMILIES OF FINITE H-COMPLEXES - REVISITED

James D. Stasheff[1]
Temple University

The most familar examples of topological groups are Lie groups. In many ways, Hopf's original introduction [6] of H-spaces was an attempt to study such groups from a non-analytic but rather homotopy theoretic point of view. Recently we have seen renewed interest in H-spaces which are finite complexes due to the discovery by Hilton and Roitberg [4] of such an H-space which was not of the homotopy type of a Lie group. Their example looked as follows: Consider the bundle $Sp(1)=S^3 \to Sp(2) \to S^7$ and let $S^3 \to E_n \to S^7$ be induced by $S^7 \xrightarrow{n} S^7$ of degree n. Hilton and Roitberg showed E_7 was an H- space.

I was subsequently [10] able to show that maps of degree 3,4,5 also induced H-spaces. For n=0, the H-space is $S^3 \times S^7$. The method I used was that of Zabrodsky [11] ; namely mixing homotopy types. The main result used was that if for each prime p a finite complex X is mod p equivalent to an H-space, then X is itself an H-space. There are various mechanisms for making this work; see Mislin's talk at this conference or Zabrodsky's original paper [11].

Curtis and Mislin [1] then constructed new H-spaces which were SU(3) bundles over S^7. Meanwhile Zabrodsky [12] showed E_2 and E_6 above were not H-spaces. This lead to classification theorems for small H-spaces [5 , 12 , 2].

It was time for a systematic approach to constructing H-spaces. Zabrodsky looked at the torsion free case while more generally Jerome Harrison, shortly before his death, looked at the situation as follows: Problem: Let $G \supset H$ be a pair of topological groups such that $G/H=S^n$ and $G \to S^n$ is a principal H-bundle classified by $\alpha \in \pi_{n-1}(H)$. For which integers k is the pullback E_K (classified by $k\alpha$) an H-space? Partial answer (Harrison [3]). Assume α of finite order and write $\alpha = \sum \alpha_p$ where α_p is of p-power order, p prime. Let $k\alpha = \sum \epsilon_p \alpha_p$. If all the ϵ_p are restricted to the set $\{-1, 0, +1\}$ then E_K is an H-space iff

1 Alfred P. Sloan Fellow. Research supported in part also by NSF grant 97836.

a) n is odd and or

b) n = 1, 3, 7

Proof: Let $X\{p\}$ denote the space obtained from X by killing all its q-primary homotopy, $q \neq p$. The following facts are relevant: For $n > 0$,

1. $S^n\{2\}$ is an H-space iff n = 1,3,7
2. $S^n\{p\}$ for $p > 2$ is an H-space iff n is odd.

Notice that $E_k\{p\} \simeq E_{\epsilon_p \alpha_p}\{p\}$

$$\simeq H\{p\} \times S^n\{p\} \qquad \text{if } \epsilon_p = 0$$

$$\simeq G\{p\} \qquad \text{if } \epsilon_p = \pm 1$$

The theorem now follows.

Using this technique and those of Curtis and Mislin, participants at the 1970 Madison Conference on Algebraic Topology went further.

Partial answer: Assume $\alpha = \sum \alpha_p$ as above and $k\alpha = \sum \epsilon_p \alpha_p$. If k is odd and for each p either

1) $\epsilon_p \not\equiv 0 \ (p)$ or
2) $\epsilon_p = 0$

then E_k is an H-space.

Meanwhile progress in the other direction was due to Zabrodsky[13] :

Partial answer (Zabrodsky): If G has no 2-torsion and E_k is an H-space, then k must be odd.

Let us look at all the examples thus provided where G, H are simple Lie groups[3] :

I. $SO(n+1)/SO(n)=S^n$. The classifying map is zero, of order 2 or of infinite order, so the above remarks do not apply. The first question which is not already decided by Hopf algebra techniques is n=6, k=3, $SO(6) \to E_3 \to S^6$.

II. $SU(n+1)/SU(n) = S^{2n+1}$. We have $\pi_{2n}(SU(n)) = Z_{n!}$ generated by this bundle. We then have H-spaces E_k for all odd k provided $n \leq 6$. E_k is not of the homotopy type of a Lie group unless $k \equiv \pm 1 (n!)$ since $\pi_{2n}(G)$ is not of odd order for G of rank n+1. The first open question is

$$\begin{array}{ccc} E_3 & \longrightarrow & SU(7) \\ \downarrow & & \downarrow \\ S^{13} & \longrightarrow & S^{13} \end{array} \quad \text{of dimension } 48.$$

III. $S_p(n+1)/S_p(n) = S^{4n+3}$. The story is much as in the unitary case. The first open question is whether E_3 is an H-space for

$$\begin{array}{ccc} E_3 & \to & Sp(4) \\ \downarrow & & \downarrow \\ S^{15} & \to & S^{15} \end{array}$$

which is of dimension 36.

IV. $G_2/SU(3) = S^6$. The above methods don't apply since $\pi_5(SU(3))=Z$. One can however show by Hopf algebra and $\mathcal{Q}(2)$- arguments that only $E_{\pm 1}$ are H-spaces. Hubbuck and Zabrodsky have shown much more. Hubbuck has shown that any finite simply connected H-complex X which rationally looks like $S^3 \times S^{11}$ must have $H^*(X;Z_2) \approx H^*(G_2;Z_2)$ while Zabrodsky's work [14] implies the same for Z_p as coefficients, $p>2$. Roitberg has shown that X need not be of the homotopy type of G_2. In general the study of finite H-complexes with torsion has barely been begun.

V. For Spin $(7)/G_2 = S^7$, the answer is easy. No new H-spaces occur since $\pi_6(G_2) = Z_3$.

VI. Spin (9)/Spin (7) = S^{15}. The classifying map is of order $2520 = 8.9.5.7$ E_k is an H-space for k odd except possibly if $k \equiv 3(9)$

A recent result of Wilkerson shows that for E_k above with $k\alpha = \sum \epsilon_p \alpha_p$, $\epsilon_p=0$ implies E_k is not homotopy equivalent to an associative H-space.

The above approach to finite H-complexes emphasizes the "p-primary parts". Recent work of Oka [9], Mimura and Toda [8] and the talks of Mislin and Curtis at this conference give particular importance to the study of (mod p) sphere bundles over spheres. In particular if $B_n(p)$ denotes the S^{2n+1} - bundle over $S^{2n+2p-1}$ classified by $\alpha_1 \in \pi_{2n+2p-2}(S^{2n+1})$, then $B_n(p)$ is an H-space mod p for $n < p$. $B_1(3)$ is mod 3 equivalent to $S_p(2)$, and $B_1(5)$ is mod 5 equivalent to G_2. For no other values of n and p is $B_n(p)$ yet known to be equivalent to an associative H-space. [[Since the conference, I have been able to show $B_n(p)$ is an H-space mod p for all $p>5$ and $B_n(5)$ is an H-space mod 5 for $n>6$.]]

As for homotopy commutativity, $S^3\{p\}$ is not for p=2 and 3 and for all p, at least is not of the homotopy type of a doubleloop space.

For a finite H-complex, rather than its p-primary parts, questions of associativity rather than of commutativity are of interest for Hubbuck [7] has shown that a homotopy commutative finite H-complex has the homotopy type of $\underbrace{S^1 \times \ldots \times S^1}_{n}$.

REFERENCES

[1] Curtis, M.; Mislin, G., Two new H-spaces, Bull AMS 76 (1970),

[2] ____________________, private communication.

[3] Harrison, J.; Stasheff, J., Families of H-spaces, QJM, to appear.

[4] Hilton, P.J., and Roitberg, J., On principal S^3-bundles over spheres, Ann. of Math. 90 (1969), 91-107.

[5] Hilton, P.J., and Roitberg, J., On the classification problem for H-spaces of rank two, Comm. Math. Helv. (1970)

[6] Hopf, H., Uber die Topologie der Gruppen-Mannigfaltigkeiten und ihre Verallgemeinerungen, Ann. of Math. (2) 42 (1941), 22 - 52.

[7] Hubbuck,J., On homotopy commutative H-spaces, Topology 8 (1969), 119-126.

[8] Mimura, M.; Toda, H., Cohomology operations and the homotopy of compact Lie groups, preprint.

[9] Oka, S., The homotopy groups of sphere bundles over spheres, J. Sci. Hiroshima U. AI 33 (1969) 161-195.

[10] Stasheff, Manifolds of the homotopy type of (non-Lie) groups, Bull. AMS 75 (1969) 998-1000.

[11] Zabrodsky, Homotopy associativity and finite CW-complexes, Topology 9 (1970), 121-128.

[12] Zabrodsky, The classification of simply connected H-spaces with three cells, I and II, preprint.

[13] Zabrodsky, On sphere extensions of classical Lie groups, Conference on Algebraic Topology 1970, Madison, Wisc.

[14] Zabrodsky, A. Secondary cohomology operations in the module of indecomposables, Summer Conference on Algebraic Topology, Aarhus, 1970

H-SPACES MOD P (I)

Guido Mislin[1]
The Ohio State University

0. INTRODUCTION

The aim of this talk is to make propaganda for the use of the localization technique in algebraic topology, especially in the study of H-spaces. We will describe the mixing technique of Zabrodsky [6] in this framework, together with several applications. Our basic theorem (2.1) states that a space X is an H-space if and only if $X_{(p)}$ (i.e. X localized at p) is an H-space for all primes p. A similar theorem holds for homotopy associative or homotopy commutative H-spaces and for loop spaces.

Using this theorem one gets (theorem 2.2) a procedure of constructing new H-spaces out of a family of spaces, which are mod certain primes H-spaces.

These theorems are applied to construct H-spaces of the Hilton-Roitberg type [4], namely spaces $X(p+1)$ such that $X(p+1) \not\simeq SU(p+1)$ but $X(p+1) \times SU(p) \simeq SU(p+1) \times SU(p)$ for all primes $p > 3$. Further we construct a (1-connected finite) H-space, which has torsion but no 2-torsion, a phenomenon which for a (1-connected) Lie group can never occure, of course. More applications of the technique of localization are given in the talk of Curtis (H-spaces mod p. II.)

1. LOCALIZATION

Let $S \subset \mathbb{Z}$ denote a multiplicatively closed set with $0 \notin S$. Then there is a functor $F : \underline{1CW.} \to \underline{1CW.}$ from the category of 1-connected pointed CW-complexes into itself, which corresponds to the localization $S^{-1} : \underline{Ab} \to \underline{Ab}$ of abelian groups in the following way:

There is a natural homotopy class $\epsilon_X : X \to FX$ inducing an isomorphism

$$S^{-1} \pi_* X \cong \pi_* FX$$

The functor F and the natural transformation ϵ may be obtained via geometric realization from the functors R_∞ an i_∞ of [2] by using the ring $R = S^{-1}\mathbb{Z}$.

---

1 This lecture represents joint work with Morton Curtis.

It is easy to see that the canonical maps $F(X \times Y) \to FX \times FY$ and $FX \vee FY \to F(X \vee Y)$ are homotopy equivalences since they induce in homotopy (resp. homology) an isomorphism. It follows that F maps H-spaces (resp. co-H-spaces) into H-spaces (resp. co-H-spaces). One may also prove that F commutes with fibrations and cofibrations up to homotopy.

If $S = \mathbb{Z} - \{0\}$ (resp. $S = \mathbb{Z} - (p)$ for a prime ideal (p)) then we write $X_{(0)}$ (resp. $X_{(p)}$) instead of FX. Note that $\pi_*(X_{(p)}) \cong (\pi_* X)_{(p)}$, the homotopy groups of X localized at p, and hence also $H_*(X_{(p)}; \mathbb{Z}) \cong H_*(X; \mathbb{Z}_{(p)}) \cong H_*(X; \mathbb{Z})_{(p)}$.

All our spaces will be in <u>1CW</u>.; for simplicity we will assume that our H-spaces have a finitely generated rational cohomology algebra and that they admit an H-space structure such that their rational cohomology be primitively generated. This guarantees that if X is an H-space then X admits an H-structure such that $\epsilon : X \to X_{(0)}$ is an H-map with respect to the standard H-structure on $X_{(0)}$ (the standard structure on $X_{(0)}$ is the one which is induced from the fact that $X_{(0)}$ has the homotopy type of a product of rational Eilenberg-MacLane spaces).

<u>1.1 Definition</u> X and Y are called p-equivalent (resp. rationally equivalent) if $X_{(p)} \simeq Y_{(p)}$ (resp. $X_{(0)} \simeq Y_{(0)}$).

Note that $X_{(p)} \simeq Y_{(p)}$ for all p does not imply that $X \simeq Y$. For instance $(E_7)_{(p)} \simeq (Sp(2))_{(p)}$ for all p but $E_7 \not\simeq Sp(2)$, where E_7 denotes the "Hilton-Roitberg criminal" [4].

<u>1.2 Definition</u> $f : X \to Y$ is called a p-equivalence (resp. a rational equivalence) if $f_{(p)}$ (resp. $f_{(0)}$) is a homotopy equivalence.

Nóte that if f is a p-equivalence for all p, then f is a homotopy equivalence. Namely $f_* : \pi_* X \to \pi_* Y$ is then a homomorphism which when localized at each maximal ideal of $\mathbb{Z}$ is an isomorphism, and hence f_* is an isomorphism by standard algebra [1; Chap.II, §3, n°3, Th.1].

An example of a p-equivalence is $\epsilon : X \to X_{(p)}$.

<u>1.3 Definition</u> X is called an H-space mod p if $X_{(p)}$ is an H-space.

For instance let $m : S^n \times S^n \to S^n$ denote a map of type $(1,2)$, n odd. Then for an odd prime p, the map

$$m_{(p)} \circ \text{can} : S^n_{(p)} \times S^n_{(p)} \simeq (S^n \times S^n)_{(p)} \to S^n_{(p)}$$

is a homotopy equivalence when restricted to a factor. Hence S^n is an H-space mod all odd primes.

2. THE MAIN THEOREM

2.1 Theorem. Let X be a finite CW complex. Then

1) X is an H-space iff X is an H-space mod p for all p.

2) X is a homotopy associative (resp. homotopy commutative) H-space if $X_{(p)}$ is a homotopy associative (resp. homotopy commutative) H-space for all p.

3) X has the homotopy type of a loop space iff $X_{(p)}$ has the homotopy type of a loop space for all p.

Sketch of proof. Of course the nontrivial part of this theorem is the passage from the "local" to the "global". Hence assume that $X_{(p)}$ is an H-space, equipped with a rationally primitive structure (so that $\epsilon : X_{(p)} \to X_{(0)}$ is an H-map with respect to the standard structure on $X_{(0)}$). Using obstruction theory one can show that in H1CW., the homotopy category of 1CW., X has the universal property of a pullback of all the maps of the $X_{(p)}$'s into $X_{(0)}$:

$$X \simeq \prod_p (X_{(p)} \to X_{(0)}) \qquad \text{in H1CW.}$$

Hence there is a natural isomorphism of functors

$$[\ ,X] \cong \prod_p ([\ ,X_{(p)}] \to [\ ,X_{(0)}]) \ : \ \text{H1CW.} \to \text{Sets.}$$

The natural transformations $[\ ,X_{(p)}] \to [\ ,X_{(0)}]$ are induced by ϵ and hence they are morphisms of monoid-valued functors. It follows that $\prod_p ([\ ,X_{(p)}] \to [\ ,X_{(0)}])$ factors through the category of monoids. Therefore $[\ ,X]$ may be factored through the category of monoids, which is equivalent to say that X admits an H-space structure. Statement 2 of the theorem is proved in the same way. For 3 we start with maps $t_p : X_{(p)} \to X_{(0)}$ between loop spaces. Since $H^*(X_{(p)}; Q)$ is an exterior algebra, all primitives lie in the image of the cohomology suspension. Hence we can deloop t_p to get maps $Bt_p : BX_{(p)} \to BX_{(0)}$. As before, we show that this family of maps has a pullback in H1CW. : $Y \simeq \prod_p (BX_{(p)} \to BX_{(0)})$. By general categorical reasons it is immediate that therefore

$$\Omega Y \simeq \prod_p (X_{(p)} \to X_{(o)})$$

Hence $X \simeq \Omega Y$.

Remark. Some finiteness conditions on X are needed to prove that $X \simeq \prod_p (X_{(p)} \to X_{(o)})$. For instance if $X = K(\prod_p \mathbb{Z}_p, n)$ then $X \not\simeq \prod_p (X_{(p)} \to X_{(o)})$.

2.2 Theorem. Let $\{p_i\}_{i \in N}$ be the set of all primes and $\{X_i\}$ a family of finite CW complexes which are all rationally equivalent. Suppose that $(X_i)_{(p_i)}$ is torsion free for large i . Then there is a finite CW complex W such that

$$W_{(p_i)} \simeq (X_i)_{(p_i)}$$

for all i .

Proof. Take $\overline{W}$ to be $\prod_i (X_i)_{(p_i)} \longrightarrow K)$ in HlCW. Here K denotes a space of the homotopy type of $(X_i)_{(o)}$. The maps $(X_i)_{(p_i)} \to K$ are $(X_i)_{(p_i)} \xrightarrow{\varepsilon} X_{(o)} \xrightarrow{\varphi_i} K$ where the φ_i's are homotopy equivalences. (The homotopy type of $\overline{W}$ will depend upon the φ_i's). Clearly $\overline{W}_{(p_i)} \simeq (X_i)_{(p_i)}$ and hence $\overline{W}$ has the homotopy type of a finite CW complex W .

Note that if in addition X_i is an H-space mod p_i for all i, then W will be an H-space by 2.1.

3. APPLICATIONS

First we want to construct loop spaces of the Harrison-Stasheff type [3]. Consider a principal bundle $H \to G \to S^n$ with H and G compact Lie groups and $n > 1$ odd. Let $\alpha \in \pi_{n-1} H$ be the characteristic element. It has finite order and can hence be written as $\alpha = \Sigma \alpha_p$ with α_p of order a power of p . Denote by E_β the total space of a principal H bundle over S^n with characteristic element $\beta \in \pi_{n-1}H$.

3.1 Corollary. If $\beta = \Sigma k_p \alpha_p$ and $(k_p, p) = 1$ for all p then E_β has the homotopy type of a loop space.

Proof. Observe that $(E_\beta)_{(p)} \simeq G_{(p)}$ for all p and apply 2.1.

As another application we construct H-spaces of the Hilton-Roitberg type [4].

3.2 Corollary. There are for all primes $p > 3$ loop spaces $X(p+1)$ such that $X(p+1) \not\simeq SU(p+1)$ but $X(p+1) \times SU(p) \simeq SU(p+1) \times SU(p)$.

Proof. Let $\alpha = \Sigma\alpha_q$ classify $SU(p) \to SU(p+1) \longrightarrow S^{2p+1}$. Then let $X(p+1) = E_\beta$ with $\beta = \sum_{q \neq p} \alpha_q + 2\alpha_p$; notice that α_p is of order p. By choosing generators x_i and y_i of $H^*(SU(p+1); \mathbb{Z}_p)$ resp. $H^*(X(p+1); \mathbb{Z}_p)$ which are mod p reduction of integral generators and such that $P_1 x_3 = m x_{2p+1}$ resp. $P_1 y_3 = n y_{2p+1}$ (deg x_i = deg y_i = i), one can see that $(m,p) = (n,p) = 1$ and $m = \pm 2n$ mod p. If $X(p+1) \simeq SU(p+1)$ then one would have $m = \pm n$ mod p; since $p > 3$ it follows that $X(p+1) \not\simeq SU(p+1)$. By Mimura and Toda [5] (see also Toda's talk at this conference), one has $SU(p+1)_{(p)} \simeq (S^5 \times S^7 \times \ldots \times S^{2p-1} \times E_\gamma)_{(p)}$ where E_γ denotes a principal S^3 bundle over S^{2p+1} with characteristic element γ of order p. Using this and Hilton-Roitberg's result $E_\gamma \times S^3 \simeq E_{2\gamma} \times S^3$ cf. [4; Cor. 3.1], one can prove that $SU(p+1) \xrightarrow{pr} S^{2p+1} \xrightarrow{\bar\beta} BSU(p)$ as well as $X(p+1) \xrightarrow{pr} S^{2p+1} \xrightarrow{\bar\alpha} BSU(p)$ are null homotopic. Therefore the corollary follows by taking the pullback of $X(p+1) \to S^{2p+1} \leftarrow SU(p+1)$. The $X(p+1)$'s are loop spaces by 3.1.

3.3 Corollary. There is a finite H-complex which has 3 torsion but no 2 torsion.

Proof. F_4 is a Lie group of type (3, 11, 15, 23) and has 2 as well as 3 torsion. Of course $F_4 \times S^7 \times S^{19}$ is an H-space mod all odd primes. By taking mod 2 the Lie group $Sp(6)$, which is rationally equivalent to $F_4 \times S^7 \times S^{19}$, we can get by 2.2 and 2.1 a finite H-complex W with $W_{(2)} \simeq Sp(6)_{(2)}$ and $W_{(p)} \simeq (F_4 \times S^7 \times S^{19})_{(p)}$ for all odd p's . Obviously W has 3 torsion but no 2 torsion.

REFERENCES

1. N. Bourbaki, Éléments de mathématique, algèbre commutative. Hermann, Paris 1961.

2. A. K. Bousfield and D. M. Kan, Homotopy with respect to a ring. (preprint).

3. J. Harrison and J. Stasheff, Families of H-spaces. To appear.

4. P. Hilton and J. Roitberg, On principal S^3 bundles over spheres. Ann. of Math. 90(1) 91-107, 1969.

5. M. Mimura and H. Toda, Cohomology operations and homotopy of compact Lie groups I. To appear in Topology.

6. A. Zabrodsky, Homotopy associativity and finite CW complexes. Topology 9(2) 121-128, 1970.

---

[1] Research supported by "Schweizerischer Nationalfonds".

H-SPACES MOD P (II)

Morton Curtis[1]
Rice University

This talk is a continuation of H-spaces mod p, I by Mislin and is a report on joint work. As remarked by Stasheff, we now know which spheres are H-spaces, which sphere bundles over spheres are H-spaces, and which spheres are H-spaces mod p. It seems natural to consider sphere bundles over spheres as H-spaces mod p. We give here some negative and some positive results on this problem, but are, of course, far from a complete answer. The key tool here, just as in part I, is localization of a space at a prime. This allows us to set up a workable obstruction theory.

1. SPHERICAL FIBRATIONS

Let q,n be odd integers greater than 1. Let F_q denote the (associative) H-space of maps $S^q \longrightarrow S^q$ of degree +1, and let BF_q be the classifying space for F_q bundles.

Lemma 1: $\pi_n(BF_q)$ is finite.

Proof: We will show $\pi_{n-1}(F_q) \cong \pi_n(BF_q)$ is finite. Let $F_q(*) \subset F_q$ consist of pointed maps. Then we have the fibration

$$F_q(*) \xrightarrow{i} F_q \xrightarrow{e} S^q$$

where i is inclusion and e is the evaluation map, $e(f) = f(*)$. Thus we have the exact sequence

$$\ldots \longrightarrow \pi_{n-1}(F_q(*)) \longrightarrow \pi_{n-1}(F_q) \longrightarrow \pi_{n-1}(S^q) \longrightarrow \ldots \quad .$$

Since q is odd and n is odd, $\pi_{n-1}(S^q)$ is finite. Since $\pi_{n-1}(F_q(*)) \simeq \pi_{n-1}(\Omega^q S^q) \simeq \pi_{n+q-1}(S^q)$, and $n>1$, this group is also finite. Thus $\pi_{n-1}(F_q)$ is finite.

<u>Lemma</u> 2: <u>Let</u> $S^q \longrightarrow X \longrightarrow S^n$ <u>be a spherical fibration</u>. <u>Then</u>

(i) $\tilde{H}^*(X;\mathbb{Q}) = \Lambda_{\mathbb{Q}}(x_q, x_n)$, <u>and</u>

(ii) $X \simeq (S^q \cup_\alpha e^n) \cup_\beta e^{n+q}$.

<u>Proof</u>: Let k be the order of the classifying map $\gamma: S^n \longrightarrow BF_q$, and let k also denote a map $S^n \longrightarrow S^n$ of degree k. Then we have

$$\begin{array}{ccccc} S^q \times S^n & \xrightarrow{f_k} & X & & \\ \downarrow & & \downarrow & & \\ S^n & \xrightarrow{k} & S^n & \xrightarrow{\gamma} & BF_q \end{array} ,$$

since $\gamma \circ k$ is nullhomotopic. Then f_k is a $\mathbb{Q}$-equivalence, proving (i).

The E_2 term of the Serre spectral sequence of this fibration is

$$H^*(S^n;\mathbb{Z}) \otimes H^*(S^q;\mathbb{Z}) .$$

Since the rational sequence collapses and there is no torsion, the integral sequence collapses, so E_2 is the cohomology of X. Thus we have the homology decomposition of X given in (ii).

The first attaching map α of this decomposition of X may be described in two ways. First, if $\chi \in \pi_{n-1}(F_q)$ is the characteristic class (adjoint of γ), then

$$\alpha = e_*\chi .$$

Secondly, if ∂ is the boundary in the homotopy exact sequence of the fibration $S^q \longrightarrow X \longrightarrow S^n$, then

$$\alpha = \partial\iota_n .$$

We recall that X has a cross section if and only if $\alpha \simeq 0$.

2. TWO MOD p LEMMAS

For most odd primes p we will see that X(p), X localized at p, is an H-space because $S^q(p)$ and $S^n(p)$ are H-spaces and

$$X(p) \simeq S^q(p) \times S^n(p).$$

Lemma 3: Let k be the order of the classifying map $\gamma \in \pi_n(BF_q)$. If p is an odd prime and $p \nmid k$, then

$$X(p) \simeq S^q(p) \times S^n(p) .$$

Proof: Referring to the diagram used in Lemma 2 we note that since p does not divide k we have $X(p) \simeq (S^q \times S^n)(p) \simeq S^q(p) \times S^n(p)$.

It follows that to check if X is an H-space mod all odd primes, only those dividing k (= order γ = order χ) need to be checked. For example, if χ is of order 2, then $X(p) \simeq S^q(p) \times S^n(p)$ for all odd primes p.

The next instance of X(p) being a product is the mod p version of a theorem of James and Whitehead [2].

Lemma 4: Let α_p be the p-primary component of α and suppose $\alpha_p = 0$. If X(p) is an H-space, then $X(p) \simeq S^q(p) \times S^n(p)$.

Proof: If $\alpha = 0$, there is an inclusion

$$S^q \vee S^n \longrightarrow X.$$

Localizing at p and using a multiplication $X(p)\times X(p) \longrightarrow X(p)$ gives an extension to $S^q(p)\times S^n(p)$ which is a homotopy equivalence.

If $\alpha \neq 0$, let ℓ be the order of α and note that $p \nmid \ell$ since $\alpha_p = 0$. Consider the pullback

$$\begin{array}{ccccc} E & \xrightarrow{f_\ell} & X & & \\ \downarrow & & \downarrow & & \\ S^n & \xrightarrow{\ell} & S^n & \xrightarrow{\gamma} & BF_q \end{array} .$$

Then E has a cross section so that we have an inclusion $i\colon S^q \vee S^n \longrightarrow E$. We localize

$$S^q \vee S^n \xrightarrow{f_\ell \circ i} X$$

at p so that the map becomes an inclusion and proceed as before.

3. THE OBSTRUCTION

Theorem 1: Let $S^q \longrightarrow X \xrightarrow{f} S^n$ be a spherical fibration with q,n odd and >1. Let

$$\mathcal{J} = \{2q,\ q+n,\ 2n,\ n+2q,\ 2n+q,\ 2n+2q\} ,$$

and let p be an odd prime. Then there exist elements of

$$\tilde{H}^i(X\wedge X;\ \pi_{i-1}(S^q;p)) \text{ for } i \in \mathcal{J}$$

such that X(p) is an H-space if these elements are zero.

Proof: Consider the diagram

$$\begin{array}{ccc}
X \vee X & \xrightarrow{\ \nabla\ } & X \\
\downarrow & & \downarrow f \\
X \times X & & S^n \\
\downarrow f\times f & & \\
S^n \times S^n & &
\end{array}$$

where ∇ is the folding map. If

$$m: S^n(p) \times S^n(p) \longrightarrow S^n(p)$$

is an H-space multiplication [1], then localizing at p gives

$$\begin{array}{ccc}
X(p) \vee X(p) & \xrightarrow{\ \nabla(p)=\nabla\ } & X(p) \\
\downarrow & \overset{\varphi}{\nearrow} & \downarrow f(p) \\
X(p) \times X(p) & \xrightarrow[\ g\ ]{} & S^n(p) \\
\downarrow f(p)\times f(p) & \overset{m}{\nearrow} & \\
S^n(p) \times S^n(p) & &
\end{array}$$

where $g = m\circ(f(p)\times f(p))$. The obstructions to the existence of an H-space structure φ lie in

$$\tilde{H}^i(X(p)\times X(p),\ X(p)\vee X(p);\ \pi_i(S^q(p))).$$

By properties of the localization functor this is

$$\tilde{H}^i((X\wedge X)(p);\ (\pi_{i-1}S^q)(p)) \simeq \tilde{H}^i(X\wedge X;\ \pi_{i-1}(S^q;p)),$$

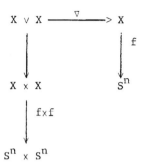

and, of course, the set $\mathcal{J}$ contains exactly the dimensions of the cells of $X \wedge X$.

For application of Theorem 1 we use the following theorem of Toda [3].

Theorem: *Let $q>1$ be odd and let p be an odd prime. Then, letting* $q = 2r+1$,

$$\pi_{q+2i(p-1)-2}(S^q;p) = \mathbb{Z}_p \qquad i = r+1,\dots,p-1$$

$$\pi_{q+2i(p-1)-1}(S^q;p) = \mathbb{Z}_p \qquad i = 1, \dots,p-1$$

$$\pi_{q+k}(S^q;p) = 0 \quad \textit{otherwise for } k < 2p(p-1)-2.$$

This easily yields a third criterion for $X(p) \simeq S^q(p) \times S^n(p)$.

Lemma 5: *Given p, if*

$$n-q < 2p(p-1)-1$$

and if there is no ℓ_p, $1 \leq \ell_p \leq p-1$, such that

$$n-q = 2\ell_p(p-1) ,$$

then $X(p) \sim S^q(p) \times S^n(p)$.

Proof: The result is automatic if $\chi_p = 0$. If not, note that $n-1$ is in the range of applicability of Toda's theorem, so we must have

$$n-1 = q+2i(p-1)-1,$$

so that such an ℓ_p exists.

Corollary: *Suppose*

$$n-q < 11 \quad \underline{\text{and}}$$
$$n-q \not\equiv 0 \ \underline{\text{mod}}\ 4.$$

Then $X(p) \sim S^q(p) \times S^n(p)$ for all odd p.

Next we note that in the range of applicability of Toda's theorem, some of the obstructions are automatically zero.

Theorem 2: Let p be a prime and suppose

$$n-q < 2p(p-1)-1 .$$

Then obstructions to $X(p)$ being an H-space lie in

$$\tilde{H}^i(X \wedge X;\ \pi_{i-1}(S^q;p))$$

for $i = q+n,\ 2n,\ 2q+2n$.

Proof: We use Theorem 1 and Lemma 5. We may assume $\chi_p \neq 0$, so that

$$n-q = 2\ell_p(p-1).$$

We claim the obstructions for cells of $X \wedge X$ of dimensions $2q, n+2q$ and $2n+q$ are zero because the coefficient groups are zero. For $2q$ we must have

$$2q-1 = q+2i(p-1)-2 \quad \text{with } i \geq r+1$$

if the coefficient group is to be nonzero. But this gives $r = i(p-1)-1$, contradicting $i \geq r+1$.

Similarly

$$n+2q-1 = q+2i(p-1)-1$$

gives $$n = (i-\ell_p)(p-1), \text{ contradicting } n \text{ being odd.}$$

The condition

$$2n+q-1 = q+2i(p-1)-1$$

gives $$n = i(p-1), \text{ contradicting } n \text{ being odd.}$$

Let $p_0, p_1, \ldots, p_t$ be (in order) the odd primes dividing the order k of χ. Then as an easy consequence of Theorem 2 we have:

<u>Theorem</u> 3: <u>If</u>

(i) $2n+q < 2p_0(p_0-1)-2$

(ii) $q+1 \not\equiv 0 \mod 2(p_i-1) \quad i = 1,\ldots,t$

(iii) $3q+1 \not\equiv 0 \mod 2(p_i-1) \quad i = 1,\ldots,t$

<u>Then</u> $X(p)$ <u>is an H-space for all odd primes</u>.

REFERENCES

[1] Adams, J. F., The sphere, considered as an H-space mod p, Quart. J. Math., Oxford (2) 12 (1962), 52-62.

[2] James, I. M., Whitehead, J. H. C., The homotopy theory of sphere bundles over spheres, I, Proc. London Math. Soc., (3) 4 (1954), 196-218.

[3] Toda, H., Composition methods in homotopy groups of spheres, Ann. of Math. Studies No. 49, Princeton.

---

[1] Research partially supported by NSF Grant GP 12715.

MOD P DECOMPOSITION OF LIE GROUPS

Hirosi Toda
Kyoto University and Northwestern University

The present talk is a summary of the following four joint works with M. Mimura and partially with R. C. O'Neill:

(1) M. Mimura, R. C. O'Neill and H. Toda, "On p-equivalence in the sense of Serre", Japanese J. of Math., to appear.

(2) M. Mimura and H. Toda, "On p-equivalences and p-universal spaces", Comm. Math. Helv., to appear.

(3) M. Mimura and H. Toda, "Cohomology operation and homotopy of compact Lie groups I", Topology Vol. 9, to appear.

(4) M. Mimura and H. Toda, "Cohomology operation and homotopy of compact Lie groups II", in preparation.

All spaces are supposed to have the same homotopy type of 1-connected finite CW-complexes.

A p-<u>equivalence</u> $f : X \longrightarrow Y$ is a map which induces isomorphisms of the mod p (co)homology groups, and

$$X \underset{p}{\xrightarrow{\sim}} Y$$

indicates the existence of such a map. Then we say that X is p-<u>decomposable</u> if for some spaces X_1, X_2, which are not p-equivalent to a point,

$$X_1 \times X_2 \underset{p}{\xrightarrow{\sim}} X \quad (\text{or } X \underset{p}{\xrightarrow{\sim}} X_1 \times X_2).$$

First of all we have to mention that the relation $\underset{p}{\xrightarrow{\sim}}$ is not an equivalence relation in general, i.e., the symmetricity does not hold. We have a sufficient condition for spaces such that the relation $\underset{p}{\xrightarrow{\sim}}$ becomes an equivalent relation.

Consider the following two conditions for a space K :

(T). For given $f : X \underset{p}{\xrightarrow{\sim}} Y$ and $g : X \longrightarrow K$ there exist $k : K \underset{p}{\xrightarrow{\sim}} K$ and $h : Y \longrightarrow K$ such that

$$\begin{array}{ccc} X & \xrightarrow{f} & Y \\ {\scriptstyle g}\downarrow & & \downarrow{\scriptstyle h} \\ K & \xrightarrow{k} & K \end{array}$$

(I). The dual condition of (T) obtained by reversing the direction of the arrows.

Theorem 1. (T) iff (I)

The proof in [1] is briefly stated as follows. Consider special cases (T') = (T) with $X = Y = S^n$ for each n and (I') = (I) with X = Y = (dim K + 1)-skeleton of K(Z,n) for each n. Then the conditions (T') and (I') are stated in words of homotopy groups and cohomology groups of K respectively. First prove (T) iff (T') and (I) iff (I'), then prove (T') iff (I') by use of n-connective fibering over K for each n.

Definition. A space K is called p-universal if it satisfies (T) or (I).

Corollary. If X or Y is p-universal then $X \xrightarrow[p]{\sim} Y$ implies $Y \xrightarrow[p]{\sim} X$.

All spaces treated here are p-universal, so the relation $\xrightarrow[p]{\sim}$ becomes an equivalent relations, denoted by $\underset{p}{\sim}$ and called a p-equivalence. Easy examples of p-universal spaces are H-spaces and co-H-spaces. In [2] we have seen that this holds for mod 0 (rational) H-spaces and mod 0 co-H-spaces, moreover we have

Theorem 2. If $H^*(X;Q) = Q[x_1,\dots,x_n]/(x_1^{a_1},\dots,x_n^{a_n})$ or X is a cofiber of a map of two suspension spaces (co-H-spaces), then X is p-universal for every p.

Corollary. 3-cell complex is p-universal for all p.

A counter example for the p-universality is the following 4-cell complex K_t. Consider the one-point union

$$L = S^3 \vee CP_2$$

of 3-sphere S^3 and complex projective plane $CP_2 = S^2 \cup e^4$. Let α_i, i = 2,3,4,5 be a generator of a free part of

$$\pi_i(L) = Z + (0 \text{ or } Z_2).$$

α_2, α_3 are given by inclusions, $\alpha_4 = [\alpha_2, \alpha_3]$ and $\alpha_5 : S^5 \longrightarrow CP_2$ is the canonical fibering. Put

$$\alpha = [[[\alpha_4,\alpha_3],\alpha_4],\alpha_3] + [[\alpha_3,\alpha_5],\alpha_5] + [[\alpha_4,\alpha_5],\alpha_4]$$

and let

$$K_t = (S^3 \vee CP_2) \cup_{t\alpha} e^{12}$$

be the mapping cone of $t\alpha \in \pi_{11}(L)$. By use of the bilinearity of the Whitehead product we have [2].

<u>Lemma</u>. <u>If</u> $f ; K_1 \longrightarrow K_1$ <u>is a rational equivalence then</u> f_* <u>is the</u> identity of $H_*(K_1;Z)$.

The identity of L can be extended to a map $g : K_t \longrightarrow K_1$ such that g is of degree t on e^{12}. If $t > 1$ then the lemma shows that there is no rational equivalence of K_1 into K_t. Thus we have

<u>Corollary</u>. K_t, $t \neq 0$, <u>is not p-universal</u>.

Now we consider the p-decomposability of compact 1-connected Lie groups G:

$$G \underset{p}{\simeq} X_1 \times X_2 \times \dots \times X_k$$

whence each X_i is mod p H-space and the p-component of the homotopy groups of G is the direct sum of those of X_i. Note that G and X_i are all p-universal.

For the case X_i = sphere, G is called <u>p-regular.</u> Kumpel-Serre theorem states that

$$G \text{ is p-regular iff } p \geq \frac{\dim G}{\operatorname{rank} G} - 1.$$

For the case X_i = sphere or $B_n(p)$, we call G <u>quasi-p-regular</u>, where $B_n(p)$ is a 3-cell complex with

$$H^*(B_n(p);Z_p) = \Lambda(x_{2n+1}, x_{2n+2p-1})$$

and

$$\mathcal{P}^1 x_{2n+1} = x_{2n+2p-1}, \qquad \deg x_j = j.$$

We may replace $B_n(p)$ by a sphere bundle over sphere with a characteristic class $\alpha_1 \in \pi_{2n+2p-1}(S^{2n+1})$ of degree p.

The following p-decompositions are given by S.Oka (J. Sci.Hiroshima Univ.(1969)) (p : odd):

$$SU(p+n) \underset{p}{\simeq} \prod_{k=1}^{n} B_k(p) \times \prod_{j=1}^{p-1} S^{2n+2j+1} \quad \text{for } 1 \leq n \leq p-1,$$

$$Sp(\tfrac{p-1}{2} + n) \underset{p}{\simeq} \prod_{k=1}^{n} B_{2k-1}(p) \times \prod_{j=1}^{(p-1)/2} S^{4n+4j-1} \quad \text{for } 1 \leq n \leq \tfrac{p-1}{2}.$$

Note that $Spin(2n+1) \underset{p}{\simeq} Sp(n)$, $Spin(2n) \underset{p}{\simeq} Sp(n-1) \times S^{2n-1}$.

An additional result is that SU(2p) and Sp(p) have similar decomposition as SU(2p-1) and Sp(p-1) by replacing $B_1(p)$ with a space $B(3,2p+1,4p-1)$ having $H^*(B(3,2p+1,4p-1);Z_p) = \Lambda(x_3, x_{2p+1}, x_{4p-1})$.

It is conjectured that SU(n) for $p \geq 3$ and Sp(n) for $p \geq 5$ are p-decomposable, but for larger values of n this is an open question.

The following theorem [3] follows from Borel-Serre's work for the classical case and from Clark's work and some additional considerations for the exceptional case.

Theorem 3. *Assume that a simple 1-connected Lie group G has no p-torsion and* $H^*(G;Z_p) = \Lambda(x_{n_1},\ldots,x_{n_\ell})$, $\deg x_{n_i} = n_i$, $n_1 < \cdots < n_\ell$, *then, up to non-zero constant,*

$$\mathcal{P}^1 x_{n_s} = x_{n_t} \quad \text{iff } n_s = n_t + 2(p-1) \text{ and } n_s \not\equiv 1 \pmod{2p-2}.$$

This implies the only if part of the following theorem [3]. The if part of the following theorem is proved by giving maps $B_n(p) \longrightarrow G$ provided some homotopy groups of G and the homotopy type of the suspension of $B_n(p)$, but the details are too long and complicated to describe here (see [3]).

Theorem 4. *G is quasi-p-regular if and only if*

$p > n/2$	for	$G = SU(n)$
$p > n$	for	$G = Sp(n)$,
$p > (n-1)/2$	for	$G = Spin(n)$,
$p > 3$	for	$G = G_2, F_4, E_6$,
$p > 7$	for	$G = E_7, E_8$

In [4] we shall prove

Theorem 5. $E_7 \underset{5}{\simeq} B(3,11,19,27,35) \times B(15,23)$

$E_7 \underset{7}{\simeq} B(3,15,27) \times B(11,23,35) \times S^{19}$

$E_8 \underset{7}{\simeq} B(3,15,27,39) \times B(23,35,47,59)$

where $B(n_1,\ldots,n_k)$ *indicates a space having* $H^*(B(n_1,\ldots,n_k);Z_p) = \Lambda(x_{n_1},\ldots,x_{n_k})$.

This finishes the problem of the p-decomposability of exceptional Lie group without p-torsion. The proof needs suitable representations of

E_7 and E_8. Note that E_6 is 3-decomposable by a work of Kumpel.

Some negative answers are known:

Theorem 6. G_2, F_4, E_6, E_7 are not 2-decomposable, and G_2 is not 3-decomposable.

This follows from Araki's work on the squaring operations in exceptional Lie groups for $p = 2$ and from $\pi_{10}(G_2) = 0$ for $p = 3$.

Finally, the following are unknown and conjectured to be not p-decomposable:

E_8	for $p = 2$,
F_4, E_7, E_8	for $p = 3$,
E_8	for $p = 5$.

ON SPHERICAL CLASSES IN THE COHOMOLOGY OF H-SPACES

Alexander Zabrodsky

The University of Illinois, Chicago

The Hebrew University, Jerusalem

0. INTRODUCTION

We consider the following problem: Given an H-space X and a mapping $f: X \to S^n$, for what integers λ does the space $X(f,\lambda)$ admit an H-structure where $X(f,\lambda)$ is the pull back in the diagram

(1)
$$\begin{array}{ccc} X(f,\lambda) & \xrightarrow{\tilde{h}} & X \\ \downarrow{\scriptstyle \tilde{f}} & & \downarrow{\scriptstyle f} \\ S^n & \xrightarrow{h_\lambda} & S^n \end{array}$$

$\deg h_\lambda = \lambda$. (i.e.: $X(f,\lambda) = \{(x,\varphi,s) \in X \times PS^n \times S^n \mid f(x) = \varphi(0)$, $h_\lambda(s) = \varphi(1)\}$ or equivalently: if f is a fibration $\tilde{f}$ is the fibration induced by h_λ).

This problem is a generalized form of the problems considered by Stasheff and Harrison [5] and that suggested by the author in [7]. In the case treated by Stasheff and Harrison in [5] some sufficient conditions for λ where given while in [8] the author gives some necessary conditions for λ under which $X(f,\lambda)$ admits an H-structure. In the latter though some restrictions on $f^*H^*(S^n,Z_2)$ had to be assumed.

This problem is strongly related to the constructions of the "new" finite CW - H-spaces originated by the Hilton-Roitberg manifold [3] and followed by Stasheff [4], Curtis and Mislin [2], and the author [6].

The main result in this study yields a necessary condition for λ (with mild restrictions on f) implied by the existence of H-structure for $X(f,\lambda)$.

To state these results one needs the following:

0.1. Definition: Let M be a graded connected algebra over $\mathcal{A}(2)$.

The $\mathcal{A}(2)$ filtration of M is given by

$$0 = F_{-1}M \subset \ldots \subset F_nM \subset F_{n+1}M \subset \ldots$$

where F_nM is the $\mathcal{A}(2)$ algebra generated by $\sum_{m \leq n} M_m$.

0.2. The Main Theorem: If $n \neq 2^j-1$, $f^*z_n \neq 0$ (where $0 \neq z_n \in H^n(S^n,Z_2)$) and $X(f,\lambda)$ admits an H-structure then λ is odd. If $n = 2^j-1$, $j > 3$, $f^*z_n \notin F_{n-1}H^*(X,Z_2)$ and $X(f,\lambda)$ admits an H-structure then λ is odd.

0.3. Corollary: (See the author's conjecture in [7]): Let $(G_n,d) = (SU(n),2)$ or $(Sp(n),4)$, $dn-1 \neq 3,7$, $f: X = G_n \to G_n/G_{n-1} = S^{dn-1}$. If $X(f,\lambda) = M(n,\lambda)$ admits an H-structure then λ is odd.

0.4 Remark: The case $dn-1 = 7$ is solved completely by Curtis-Mislin in [2] $(d = 2)$ and by the author in [7] $(d = 4)$. The necessary (and in these instances - sufficient) conditions for λ are λ = odd or $\lambda \equiv d$ mod the order of $\pi_6(G_n)$.

The main theorem follows from the following theorems:

0.5. Theorem 1: Let X be an H-space, $f: X \to S^n$. If $n \neq 2^j-1$, $0 \neq z_n \in H^n(S^n,Z_2)$ then $f^*z_n \in F_{n-1}H^*(X,Z_2)$.

As a consequence of theorem 1 one gets:

0.6. Theorem 2: Let X , f , $X(f,\lambda)$ be as in (1). Assume $n \neq 2^j-1$ and $f^*z_n \neq 0$ $(0 \neq z_n \in H^n(S^n,Z_2))$. If $X(f,\lambda)$ admits an H-structure then λ is odd.

0.7. Definition: Let Y be a CW-complex. A class $y \in H^n(Y,Z_2)$ is said to be detected by a secondary operation if there exists a GEM K, $K = \prod_j K(Z_2,n_j)$, $n_j < n$ and a mapping $g: Y \to K$ so that $j^*y \in F_{n-1}H^*(\tilde{Y},Z_2)$, where $j: \tilde{Y} \to Y$ is the fiber of g (i.e., $j: \tilde{Y} \to Y$ is the fibration induced by g from the path-fibration $\Omega K \to \mathcal{L}K \to K$).

One can easily see that if y can be detected by a secondary operation one can always choose K and g to satisfy $\operatorname{im} g^* = F_{n-1}H^*(Y,Z_2)$.

<u>0.8. Theorem 3:</u> Let X be an H-space, $f\colon X \to S^n$, $n = 2^j-1$, $j > 3$. Then f^*z_n can be detected by a secondary operation.

From theorem 3 one gets:

<u>0.9. Theorem 4:</u> Let X, f, $X(f,\lambda)$ be as in (1). If $f^*z_n \notin F_{n-1}H^*(X,Z_2)$, $n = 2^j-1$, $j > 3$ and $X(f,\lambda)$ admits an H-structure then λ is odd.

The main theorem is composed of theorems 2 and 4.

<u>1. Proof of Theorems 1 and 2.</u>

<u>1.1. Definition:</u> Let X,μ be an H-space, G - a group, $f\colon X \to G$. The H-deviation of f is the mapping $D_f\colon X \wedge X \to G$ given by

$$D_f(x,y) = f(\mu(x,y)) \cdot f(y)^{-1} \cdot f(x)^{-1} .$$

A similar definition for D_f preserving all its desired properties can be given assuming only that G is a homotopy associative H-space with a homotopy multiplicative inverse $c\colon G \to G$.

It is obvious that

<u>1.2. Lemma:</u> f is an H-space if and only if $D_f \sim *$.

<u>1.3. Lemma:</u> Let $h\colon X' \to X$ and $g\colon G \to G'$ be H-mappings (G,G' - homotopy associative H-spaces with homotopy multiplicative inverse. Let $f\colon X \to G$. Then:

(a) $D_{g\circ f} \sim g\circ D_f$

(b) $D_{f\circ h} \sim D_f \circ h \wedge h$

We also need the following:

<u>1.4. Lemma:</u> Consider the commutative diagram of spaces and maps:

$$\begin{array}{ccc} X_1 \wedge X_2 & \xrightarrow{\ h_1\ } & B \\ \big\downarrow {\scriptstyle f_1 \wedge f_2} & & \big\downarrow {\scriptstyle f} \\ Y_1 \wedge Y_2 & \xrightarrow{\ h_2\ } & B' \end{array}$$

Let $r: E \to B$, $r_i: E_i \to X_i$ $(i = 1,2)$ be the fibrations induced by f and f_i from the path fibrations $\Omega B' \to \mathcal{L} B' \to B'$ and $\Omega Y_i \to \mathcal{L} Y_i \to Y_i$ respectively. Then there exists a (homotopy) commutative diagram of liftings:

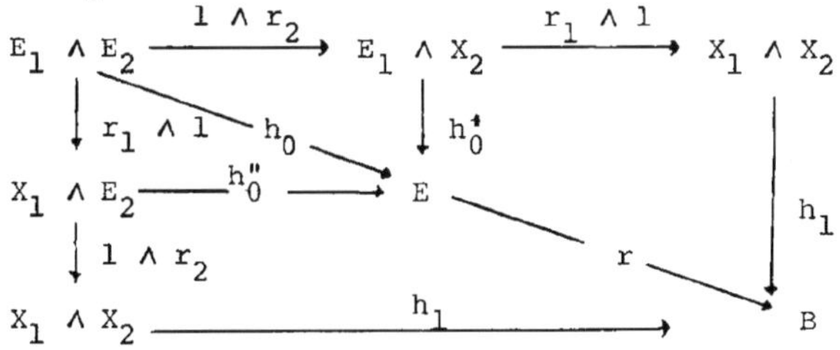

<u>Proof:</u> One has:

$$E_i = \{(x,\varphi) \in X_i \times \mathcal{L} Y_i \mid f_i(x) = \varphi(1)\}$$

$$E = \{(b,\lambda) \in B \times \mathcal{L} B' \mid f(b) = \lambda(1)\}$$

Define:

$$h_0'((x_1,\varphi_1),x_2) = h_1(x_1,x_2), \mathcal{L}h_2(\varphi_1,f_2(x_2))$$

$$h_0''(x_1,(x_2,\varphi_2)) = h_1(x_1,x_2), \mathcal{L}h_2(f_1(x_1),\varphi_2)$$

(note that $(\varphi_1,f_2(x_2))$ and $(f_1(x_1),\varphi_2)$ define elements in $\mathcal{L}(Y_1 \wedge Y_2))$ and put

$$h_0((x_1,\varphi_1),(x_2,\varphi_2)) = h_1(x_1,x_2) \ , \mathcal{L} h_2(\varphi_1,\varphi_2) \ .$$

One can see that $h_0 \sim h_0'(1 \wedge r_2) \sim h_0''(r_1 \wedge 1)$.

<u>1.5. Proposition:</u> Let X,μ be a k-connected H-space, $g: X \to K = \prod_j K(Z_2,n_j)$ $(k < n_j < n)$ an H-mapping with $\operatorname{im} g^* = F_{n-1}H^*(X,Z_2)$. (See remark at the end of 0.7). Let $j: \tilde{X} \to X$ be the fiber of g. Given any (generalized) 2-stage Postnikov system $E: \Omega K_1 \overset{j_0}{\to} E \overset{r_0}{\to} K_0$ (r_0 - a fibration and a loop map), $K_0 = \prod_s K(Z_2 m_s)$, $K_1 = \prod_t K(Z_2,\ell_t)$,

$m_s \leq n+k$, $\ell_t \leq 2n$. Then for every $f: X \to E$ $f \circ j$ is an H-mapping.

<u>Proof:</u> By 1.3(a) $D_{r_0 f} = r_0 D_f: X \wedge X \to K_0$. As $m_s \leq n+k$ one has the following factorization of D_f:

$$\begin{array}{ccc} X \wedge X & \xrightarrow{\ D_f\ } & E \\ \Big\downarrow{\scriptstyle g \wedge g} & & \Big\downarrow{\scriptstyle r_0} \\ K \wedge K & \longrightarrow & K_0 \end{array}$$

By 1.4 there exists

$$D_1\colon \tilde{X} \wedge X \to \Omega K_1$$
$$D_2\colon X \wedge \tilde{X} \to \Omega K_1 \qquad \text{so that}$$

$D_1(1 \wedge j) \sim D_2 \circ (j \wedge 1)$, $j_0 D_1 \sim D_f(j \wedge 1)$, $j_0 \circ D_2 \sim D_f(1 \wedge j)$.

Now, for every t the fundamental class $\sigma^* \iota_t \in H^{\ell_t - 1}(\Omega K_1, Z_2)$ satisfies

$$(1 \otimes j^*) D_1^* \sigma^* \iota_t = (j^* \otimes 1) D_2^* \sigma^* \iota_t \in [\overline{\text{im}}\ j^* \otimes \bar{H}^*(\tilde{X}, Z_2)] \cap$$
$$\cap\ [\bar{H}^*(\tilde{X}, Z_2) \otimes \overline{\text{im}}\ j^*] = \overline{\text{im}}\ j^* \otimes \overline{\text{im}}\ j^*$$

As $\ell_t \leq 2n$ and im j^* is $n-1$ connected, it follows that $(1 \otimes j^*) D_1^* \sigma^* \iota_t = 0$, $* \sim D_1(1 \wedge j)$, $* \sim j_0 \circ D_1 \circ (1 \wedge j) \sim D_f(j \wedge 1)(1 \wedge j) =$

$$= D_f(j \wedge j).$$

Applying 1.3(b) - $D_{f \circ j} \sim *$ and by 1.2 $f \circ j$ is an H-mapping.

<u>1.6. Proof of Theorem 1:</u> As $n \neq 2^j - 1$

$$Sq^{n+1} = \sum_i \alpha_i \gamma_i \ , \ \alpha_i, \gamma_i \in \mathcal{Q}(2) .$$

Let $h_0\colon K_0 = K(Z_2, n) \to \prod_i K(Z_2, n+|\gamma_i|) = K_1$, $h_0^* \iota_{n+|\gamma_i|} = \gamma_i \iota_n$. Let $g_0\colon S^n \to K(Z_2, n)$, $g_0^* \iota_n = z_n$ and $\Omega K_1 \xrightarrow{j_1} E_1 \xrightarrow{r_1} K_0$ be the fibration induced by h_0 from the path fibration $\Omega K_1 \to \mathcal{L} K_1 \to K_1$. Then g_0 lifts to $g_1\colon S^n \to E_1$. There exists $\omega \in H^{2n}(E_1, Z_2)$, $j_1^* \omega = \sum \alpha_i \sigma^* \iota_{n+|\gamma_i|}$ and $\mu_{E_1}^* \omega = \omega \otimes 1 + 1 \otimes \omega + r_1^* \iota_n \otimes r_1^* \iota_n$ (μ_{E_1} - the loop multiplication in E_1). Let $f_0 = g_0 \circ f$, $f_1 = g_1 \circ f$. As the hypothesis of 1.5 holds $f_1 \circ j\colon \tilde{X} \to E_1$ is an H-mapping. Let $\tilde{\mu}$ be the multiplication in $\tilde{X}$. As $g_1^* \omega = 0$, $f_1^* \omega = 0$, $0 = \tilde{\mu}^* j^* f_1^* \omega =$ $j^* f_1^* \omega \otimes 1 + 1 \otimes j^* f_1^* \omega + j^* f_1^* r_1^* \iota_n \otimes j^* f_1^* r_1^* \iota_n = j^* x \otimes j^* x$.

Hence, $j^* x = 0$ and as ker j^* is the ideal generated by

$F_{n-1}H^*(X,Z_2)$, $x \in F_{n-1}H^*(X,Z_2)$.

1.7. Proof of Theorem 2: Let $f: X \to S^n$, $f^*z_n = x \neq 0$.

Consider the diagram:

$$\begin{array}{ccccc} & & X(f,\lambda) & \xrightarrow{\tilde{h}} & X \\ & \overset{\tilde{i}}{\nearrow} & \downarrow \tilde{f} & & \downarrow f \\ F & \xrightarrow{i} & S^n & \xrightarrow{h_\lambda} & S^n \end{array}$$

where $i: F \to S^n$ and $\tilde{i}: F \to X(f,\lambda)$ are the fibers of h_λ and $\tilde{h}$ respectively. If λ is even

$$H^*(F,Z_2) \approx H^*(S^n,Z_2) \otimes H^*(\Omega S^n,Z_2) = \Lambda(y_n) \otimes \Lambda(y_{n-1},y_{2n-2},\ldots),$$

$y_n = i^*z_n$.

As $f^*z_n \neq 0$ a simple Serre spectral sequence argument implies that $\tilde{h}^*: H^*(X,Z_2) \to H^*(X(f,\lambda),Z_2)$ is an isomorphism in $\dim < n$, hence, $F_{n-1}\tilde{h}^*: F_{n-1}H^*(X,Z_2) \to F_{n-1}H^*(X(f,\lambda),Z_2)$ is onto. As $i^*z_n = \tilde{i}^*\tilde{f}^*z_n \neq 0$, $\tilde{f}^*z_n \in \operatorname{im} \tilde{h}^*$, $\tilde{f}^*z_n \notin F_{n-1}H^*(X(f,\lambda),Z_2)$ and by theorem 1 $X(f,\lambda)$ does not admit an H-structure.

2. PROOF OF THEOREMS 3 AND 4.

2.1. Proof of Theorem 3: By the classical work of Adams [1] there exists a (generalized) two stage Postnikov system

$$\begin{array}{ccc} \Omega K_1^{(m)} & \longrightarrow & E_1^{(m)} \\ & & \downarrow r^{(m)} \\ & & K(Z_2,m) \xrightarrow{h_0} \prod_{t<j} K(Z_2,m+2^t) = K_1^{(m)} \end{array}$$

$h_0^*\iota_{m+2^t} = Sq^{2^t}\iota_m$ and classes $v_s^{(m)} \in H^*(E_1^{(m)},Z_2)$, $\alpha_s \in \mathcal{A}(2)$ so that $\sum_s \alpha_s v_s^{(m)} = r^{(m)*}Sq^{2^j}\iota_m$. For $m = 2^j-1$ write $K(Z_2,m) = K_0$, $E_1^{(m)} = E_1$, $r^{(m)} = r_1$, $K_1^{(m)} = K_1$ and $v_s^{(m)} = v_s$.

Consider the 3-stage Postnikov system

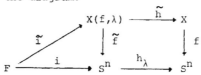

$$\begin{array}{ccccc}
\Omega K_2 & \xrightarrow{j_2} & E_2 & & \\
 & & \downarrow r_2 & & \\
 & & E_1 & \xrightarrow{h_2} & K_2 = \prod_s K(Z_2,|v_s|) \\
 & & \downarrow r_1 & & \\
 & & K_0 & \longrightarrow & K_1
\end{array}$$

$h_2^* \iota_{|v_s|} = v_s$. Then there exists a class $w \in H^{2^{j+1}-2}(E_2,Z_2)$, $j_2^* w = \sum_s \alpha_s \sigma^* \iota_{|v_s|}$ and $\mu_{E_2}^* w = w \otimes 1 + 1 \otimes w + r_2^* r_1^* \iota_{2^j-1} \otimes r_2^* r_1^* \iota_{2^j-1}$. As in the proof of theorem 1 one has a factorization:

$$\begin{array}{ccccc}
 & & & \xrightarrow{g_2} & E_2 \\
 & & & & \downarrow \\
 & & & \xrightarrow{g_1} & E_1 \\
 & & & & \downarrow \\
X & \xrightarrow{f} & S^n & \xrightarrow{g_0} & K_0
\end{array}$$

$g_0^* \iota_{2^j-1} = z_{2^j-1}$. Put $f_i = g_i \circ f$. Obvously, $f_2^* w = 0$.

As E_1 satisfies the hypothesis of E_1 in 1.5 $f_1 \circ j$ is an H-mapping. Now, as $r_2 \circ D_{f_2 \circ j} \sim D_{f_1 \circ j} \sim *$ $D_{f_2 \circ j}: \widetilde{X} \wedge \widetilde{X} \to E_2$ can be lifted to $D_1: \widetilde{X} \wedge \widetilde{X} \to \Omega K_2$ and as $\pi_m(\Omega K_2) = 0$ for $m \geq 2(2^j-1)$ $D_{f_2 \circ j}^* \omega \in F_{2^j-2} H^*(\widetilde{X},Z_2) \otimes H^*(\widetilde{X},Z_2) + H^*(\widetilde{X},Z_2) \otimes F_{2^j-2} H^*(\widetilde{X},Z_2)$. By the definition of $D_{f_2 \circ j}$ one has:

$D_{f_2 \circ j} \circ \Lambda = \mu_{E_2}[f_2 \circ j \circ \widetilde{\mu} \times \mu_{E_2}(c \circ f_2 \circ j \times c \circ f_2 \circ j)] \circ \Delta_2$ where

$\Lambda: X \times X \to X \wedge X$ is the identification map,

$\Delta_2: X \times X \to (X \times X) \times (X \times X)$ is the diagonal

μ_{E_2} and $\widetilde{\mu}$ - the multiplications in $\widetilde{X}$ and E_2 respectively,

$c: E_1 \to E_2$ the multiplicative inverse.

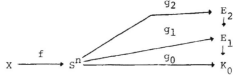

As $c^* r_2^* r_1^* \iota_{2^j-1} = r_2^* r_1^* \iota_{2^j-1}$, $c^* w = w + (r_2^* r_1^* \iota_{2^j-1})^2$, $j^* x$ - is primitive and $f_2^* w = 0$ one can compute to obtain $D_{f_2 \circ j}^* w =$ $j^* x \otimes j^* x \in F_{2^j-2} H^*(\widetilde{X},Z_2) \otimes H^*(\widetilde{X},Z_2) + H^*(\widetilde{X},Z_2) \otimes F_{2^j-2} H^*(\widetilde{X},Z_2)$ and hence, $j^*x \in F_{2^j-2}H^*(\tilde{X},Z_2)$.

<u>2.2. Proof of Theorem 4:</u> Let $j\colon \tilde{X} \to X$ and $j'\colon \tilde{X}(f,\lambda) \to X(f,\lambda)$ be the fibers of $g\colon X \to K = \prod_j K(Z_2,n_j)$ and $g'\colon X(f,\lambda) \to K' = \prod_{j'} K(Z_2,m_j)$ respectively: $\operatorname{im} g^* = F_{2^j-2}H^*(X,Z_2)$, $\operatorname{im} g'^* = F_{2^j-2}H^*(X(f,\lambda),Z_2)$.
If λ is even, as in the proof of theorem 2 $\tilde{h}\colon X(f,\lambda) \to X$ yields an isomorphism of Z_2-cohomology in $\dim < 2^j-1$. Hence, we may assume that $K = K'$ and $g' = g\circ\tilde{h}$ and one gets a diagram similar to that in the proof of theorem 2:

$$\begin{array}{ccccc} & \overset{\tilde{i}\circ j}{\nearrow} & \tilde{X}(f,\lambda) & \longrightarrow & \tilde{X} \\ F & & \downarrow{\scriptstyle \tilde{f}\circ j} & & \downarrow{\scriptstyle f\circ j} \\ & \underset{i}{\searrow} & S^n & \xrightarrow{h_\lambda} & S^n \end{array}$$

The hypothesis $f^*z_n \notin F_{2^j-2}H^*(X,Z_2)$ yields $(f\circ j)^*z_n \neq 0$ and following the steps of the proof of theorem 2, theorem 4 follows.

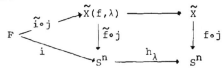

REFERENCES

[1] Adams, J.F., On the nonexistence of elements of Hopf invariant one, Ann. of Math. 72, (1960), pp.20-104.

[2] Curtis, M. and Mislin, G., Two new H-spaces, Bull. Amer. Math. Soc. 4, (1970), p.851.

[3] Hilton, P. and Roitberg, J., On principal S^3-bundles over spheres, Ann. of Math. 90, (1969), pp.91-107.

[4] Stasheff, J., Manifolds of the homotopy type of (non Lie) groups, Bull. Amer. Math. Soc. 75, (1969), pp.998-1000.

[5] Stasheff, J., and Harrison J., Families of H-spaces, Mimeographed.

[6] Zabrodsky, A., Homotopy associativity and finite CW-complexes, Topology 9, (1970), pp.121-128.

[7] ____________, The Classification of simply connected H-spaces with three cells I, Memeographed.

[8] ____________, On sphere extension of classical groups, transactions of the summer conference on algebraic topology, Madison, Wisconsin, 1970.

ON THE HOMOTOPY CLASSIFICATION OF RANK 3-H-SPACES

Joseph Roitberg*
State University of New York at Stony Brook

In this paper, we announce some results on the homotopy classification of rank 3 H-spaces. The rank 2 case has been dealt with in [6], [7], [14] and in the present work, an attempt is made to combine the various techniques used in these papers to study the rank 3 case.

The material discussed here is part of work being carried out in collaboration with G. Mislin; a detailed account will appear in a forthcoming joint publication [1].

§1. We make the blanket assumption that all spaces considered in this paper are finite-dimensional, 1-connected and (homologically) torsion-free. Under slightly less stringent conditions (replace torsion-free by 2-torsion-free), Hubbuck [9] has enumerated all possible types of primitively generated rank 3 H-spaces. In the present paper, we limit ourselves to discussion of two of these types, namely (3,5,7) and (3,7,11), of which the prototypes are SU(4) and Sp(3) respectively. The other cases will be treated in [11]. (We remark that recent unpublished work of Hubbuck enables one to relax the 2-torsion-free assumption somewhat. For example, in the (3,5,7) case, this assumption can be dropped completely. However, as the example $G_2 \times S^7$ (of type (3,7,11)) shows, the assumption is not redundant.)

We proceed now to discuss our techniques and state our main results.

We take the viewpoint that a finite-dimensional H-space X "behaves" like a parallelizable smooth manifold [2], [7]. More precisely, X is a Poincare space which is stably reducible, or equivalently, the

---

*Based on a talk delivered by the author at the Conference on H-spaces, Neuchatel, August, 1970.

Spivak normal spherical fibration $\nu(X)$ of X is fibre homotopy trivial [10]. This explains the relevance of the following lemma, generalizing a result of Casson [4].

Lemma. Let X be a 1-connected Poincare space and let $h : X \to S^n$ $(n \geq 2)$ be a fibration with 1-connected fibre F homotopy equivalent to a finite complex. Then: (a) F is a Poincare space, and (b) $i : F \to X$ is the inclusion, $i^*\nu(X) = \nu(F)$.

Remarks. (1) Even if X has a model M which is a closed, smooth manifold, M need not submerge onto S^n, even in the topological sense [8], [12].

(2) If X is an H-space, so that $\nu(X)$ is trivial, the lemma implies that $\nu(F)$ is also trivial. F itself might be expected to be an H-space but this is not so in general, as the following example† shows. Let $\Pi \vee \eta^2 : SU_3 \vee S^7 \to S^5$ be the map which is the natural projection Π on the first summand and the essential map η^2 on the second summand. Then it can be shown [11] that $\Pi \vee \eta^2$ extends to a map $h : SU_3 \times S^7 \to S^5$ and that the "fibre" F of h is homotopy equivalent to a cell complex $S^3 \cup e^7 \cup e^{10}$ with $\Pi_6 F = Z_6$. But a theorem of Zabrodsky [14] implies that such an F cannot be an H-space.

To apply the lemma to an H-space of type (3,5,7) (resp. (3,7,11)), we construct a map $h : X \to S^n$ with $n = 7$ (resp. $n = 11$) which induces an isomorphism on n-dimensional cohomology. Such an h can then be shown to have "fibre" F homotopy equivalent to a cell complex of the form $S^3 \cup e^5 \cup e^8$ (resp. $S^3 \cup e^7 \cup e^{10}$) and the lemma is then used to conclude that these complexes are precisely the ones studied in the rank 2 case.

To produce the appropriate map $h : X \to S^n$, we proceed by obstruction theory. Essential use is made of the behaviour of certain secondary cohomology operations on the cohomology of X and on its

†This example was produced by Mislin in response to a question raised at the conference.

projective plane P_2X [3], and also of the Browder-Spanier duality theorem (for the last obstruction). We thus have

Theorem 1. *If X is an H-space of type (3,5,7), then X fibres over S^7 with fibre homotopy equivalent to SU_3.*

Theorem 2. *If X is an H-space of type (3,7,11), then X fibres over S^{11} with fibre homotopy equivalent to an orthogonal S^3-bundle over S^7.*

§2. In this section we indicate how Theorems 1 and 2 can be refined. We dirst discuss a general result, based on recent work of Arkowitz [1].

Let X be a finite H-complex of type $(r_1,\ldots,r_s)$, let $n_1 \leq \ldots \leq n_k$ denote some subsequence of $r_1 \leq \ldots \leq r_s$, and let $N = \sum_{i=1}^{k} n_i$. Denoting a generator of $H_{n_i}X$ by ξ_{n_i} consider the N-cell of (a homology decomposition of) X which determines the Pontryagin product $\xi_{n_1}*\ldots*\xi_{n_k}$; call the attaching map of this N-cell ω. Finally, let ℓ_{n_i} denote the index of $h(\Pi_{n_i}X)$ in $H_{n_i}X$, h the Hurewicz homomorphism, and let $\ell = \prod_{i=1}^{k}\ell_{n_i}$.

Theorem 3. *There exist elements $\alpha_{n_i} \in \Pi_{n_i}X^{N-1}$ $(i = 1,\ldots,k)$ such that $\pm\ell\omega \in [\alpha_{n_1},\ldots,\alpha_{n_k}]$, where $[\alpha_{n_1},\ldots,\alpha_{n_k}]$ is the (higher order) Whithead product set. More precisely, $\pm\ell\omega = \theta(h(\alpha_{n_1})*\ldots*h(\alpha_{n_k}))$, where θ is the composition*

$$H_NX \rightarrow H_N(X,X^{N-1}) \overset{h}{\underset{\approx}{\leftarrow}} \Pi_N(X,X^{N-1}) \rightarrow \Pi_{N-1}X^{N-1}.$$

We now apply Theorem 3 to our stituation. Note that Theorem 1 provides a weak converse to a theorem of Curtis-Mislin [5], asserting that any principal SU(3)-bundle over S^7 is an H-space. Although we cannot prove that an H-space X, of type (3,5,7), must be homotopy equivalent to a principal SU(3)-bundle over S^7, we have the following partial result, using Theorem 3.

Theorem 4. If X is an H-space of type (3,5,7), then the 10-skeleton of X is homotopy equivalent to the 10-skeleton of one of the four homotopy types of principal SU(3)-bundles over S^7.

Next, we return to the (3,7,11) case and try to strengthen the conclusion of Theorem 2. In fact, we may prove, once again with the help of Theorem 3, the following.

Theorem 5. If X is an H-space of type (3,7,11), then the fibre F of the map $h : X \to S^{11}$ of Theorem 2 is homotopy equivalent to a principal S^3-bundle over S^7.

This theorem can also be proved without using Theorem 3, by relying instead on localization techniques.

§3. The conclusion of Theorem 5 can be interpreted as a statement about the second attaching map β of the Fibre $F \simeq S^3 \cup_\alpha e^7 \cup_\beta e^{10}$. It is natural to enquire whether or not there are any restrictions on the first attaching map α. Now $\alpha \in \Pi_6 S^3 = Z_{12}$ and so $\alpha = k\omega$, ω the Blakers-Massey generator. By using Zabrodsky's mixing technique [13] one may show

Theorem 6. If k is odd, then $E_{k\omega}$, the principal S^3-bundle over S^7 which is classified by $k\omega$, occurs as the fibre of a suitable fibration $h : X \to S^{11}$, X an H-space of type (3,7,11).

On the other hand, if k is even, then mixing the "well-known" (3,7,11) H-spaces will not yield an H-space with fibre $E_{k\omega}$. We are thus led to pose the

Conjecture. If k is even, then $E_{k\omega}$ does not occur as the fibre of a fibration $h : X \to S^{11}$ with X an H-space of type (3,7,11).

We believe that a study of the K-theory of the projective plane of a (3,7,11) H-space will enable us to answer this conjecture affirmatively. It is hoped that this conjecture will appear as a theorem in [11].

REFERENCES

[1] M. Arkowitz, A homological method for computing certain Whitehead products, Bull. Amer. Math. Soc. 74 (1968) 1079-1082.

[2] W. Browder and E. Spanier, H-spaces and duality, Pacific J. Math. 12 (1962) 411-414.

[3] W. Browder and E. Thomas, On the projective plane of an H-space, Ill. J. Math. 7 (1963) 492-502.

[4] A. Casson, Fibrations over spheres, Topology 6 (1967) 489-499.

[5] M. Curtis and G. Mislin, Two new H-spaces, Bull. Amer. Math. Soc. 76 (1970) 851-852.

[6] M. Curtis and G. Mislin, (to appear).

[7] P. Hilton and J. Roitberg, On the classification problem for H-spaces of rank two, Comm. Math. Helv. (to appear).

[8] P. Hilton and J. Roitberg, Note on quasifibrations and fibre bundles, Ill. J. Math. (to appear).

[9] J. Hubbuck, Generalized cohomology operations and H-spaces of low rank, Trans. Amer. Math. Soc. 141 (1969) 335-360.

[10] N. Levitt, Normal fibrations for complexes, Ill. J. Math. 14 (1970) 385-408.

[11] G. Mislin and J. Roitberg, Remarks on the homotopy classification of finite-dimensional H-spaces, (in preparation).

[12] W. Sutherland, Homotopy-smooth sphere fibrings, Bol. Soc. Mat. Mex. (1966) 73-79.

[13] A. Zabrodsky, Homotopy associativity and finite CW complexes, Topology 9 (1970) 121-128.

[14] A. Zabrodsky, The classification of simply-connected H-spaces with three cells, I and II, (to appear).

H-MAPS BETWEEN SPHERES

François Sigrist

Université de Neuchâtel (Suisse)

Using a result of Barratt, Curjel and Arkowitz [1] show that the essential map from S^7 to S^3 is not an H-map with respect to the usual multiplications on both spheres. The question arises naturally whether this map could become an H-map with respect to suitably chosen exotic multiplications; we shall see that this is not the case.

Let X and Y be H-spaces. We use the expression "$f:X\to Y$ is an H-map" to denote the following: there exist multiplications on X and Y such that f is an H-map with respect to them. We further deliberately confuse maps and homotopy classes. Between S^1, S^3, and S^7, non-trivial maps are multiples of ι_1, ι_3, ι_7, and $\alpha: S^7 \to S^3$.

Recall that $\pi_7(S^3)=Z_2$, and that both compositions $\nu'\circ\eta_6$ and $\eta_3\circ\nu_4$ can be taken as generators. We prove:

Theorem: 1) $n\iota_1$ is always an H-map

2) $n\iota_3$ is an H-map iff $\nu_2(n) \neq 1$ or 2

3) $n\iota_7$ is an H-map iff $\nu_2(n) \neq 1, 2$, or 3

4) α is not an H-map

($\nu_2(n)$ is the exponent of 2 in the decomposition of n into prime powers)

Proof: Part (1) is trivial because S^1 is a commutative H-space. Ingredients for the proof of (2) are collected and detailed in Curjel-Arkowitz [2] ; we briefly give the arguments:

a) The group $\pi(S^3\times S^3, S^3)$ is nilpotent of class 3 (3-fold commutators vanish); its commutator subgroup is cyclic of order 12, generated by the commutator map.

b) multiplications on S^3 can be obtained from the quaternionic multiplication in defining $m_r(x,y)=xy[x,y]^r$. m_s is homotopic to m_r iff $r \equiv s (\mathrm{mod}\ 12)$.

c) the map $n:S^3 \to S^3$ is an H-map iff there exist r and s such that $m_s \circ (n \times n) = n \circ m_r$ in $\pi(S^3 \times S^3, S^3)$. This equation reduces, using the nilpotency of this group, to $n^2(2s+1) \equiv n(2r+1) \pmod{24}$. The existence of r and s implies easily $\nu_2(n) \neq 1, 2$, and conversely, proving (2).

The proof of (3) is similar. S^7 is not homotopy-associative, but $\pi(S^7 \times S^7, S^7)$ is known to be a group, and arguments a), b), and c) apply.

To prove (4), suppose $\alpha: S^7 \to S^3$ is an H-map. Then there exist m and m' such that the following diagram commutes:

$$\begin{array}{ccc} S^7 \times S^7 & \xrightarrow{\ m\ } & S^7 \\ {\scriptstyle \alpha \times \alpha}\downarrow & & \downarrow{\scriptstyle \alpha} \\ S^3 \times S^3 & \xrightarrow{\ m'\ } & S^3 \end{array}$$

Apply then the Hopf construction to the diagram, getting the commutative diagram

$$\begin{array}{ccc} S^{15} & \xrightarrow{\ H(m)\ } & S^8 \\ {\scriptstyle \alpha * \alpha}\downarrow & & \downarrow{\scriptstyle \Sigma\alpha} \\ S^7 & \xrightarrow{\ H(m')\ } & S^4 \end{array}$$

H(m) and H(m') are Hopf constructions on multiplications, therefore elements of Hopf invariant one. In particular, $H(m) = \sigma_8 + k\Sigma\delta$ in $\pi_{15}(S^8)$, for some $\delta \in \pi_{14}(S^7)$. Now recall $\Sigma\alpha = \eta_4 \circ \nu_5 = \Sigma\nu' \circ \eta_7$. It is easily seen that $\eta_7 \circ \Sigma\delta$ is zero. But $\eta_4 \circ \nu_5 \circ \sigma_8$ is non-zero: This can be tracked through Toda's book, and is explicitly given in a paper by Mimura-Toda [3]: $\eta_4 \circ \nu_5 \circ \sigma_8 = \Sigma(\varepsilon_3 \circ \nu_{11} + \nu' \circ \varepsilon_6)$. The composition $\Sigma\alpha \circ H(m)$ is consequently non-trivial in $\pi_{15}(S^4)$. On the other hand, $H(m') \circ (\alpha * \alpha)$ is zero: by the Barratt-Hilton formula, $\alpha * \alpha$ is already zero. The Hopf construction diagram cannot be commutative, a contradiction proving (4).

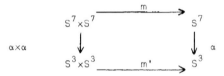

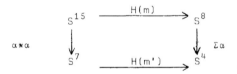

REFRENCES

[1] ARKOWITZ-CURJEL: On maps of H-spaces
Topology 6 (1967) 137-148

[2] ARKOWITZ-CURJEL: Some properties of the exotic multiplications on the three-sphere.
Quat. J. Math. Oxford (2), 20 (1969), 171-176

[3] MIMURA-TODA: Homotopy groups of SU(3), SU(4) and Sp(2).
J. Math. Kyoto Univ. 3-2(1964) 217-250

ON H-SPACES OF FINITE DIMENSION

C. R. CURJEL and R. R. DOUGLAS*

Let $\underline{CW}_0$ be the homotopy category of pointed connected finite CW-complexes.

Theorem: Up to homotopy type there are only finitely many objects in $\underline{CW}_0$ of dimension $\leq N$ which admit a multiplication.

Corollary: Up to isomorphism of group objects in $\underline{CW}_0$ there are only finitely many group objects in $\underline{CW}_0$ whose underlying space is an object of $\underline{CW}_0$ of dimension $\leq N$. - The corollary does not hold if "group in $\underline{CW}_0$" is replaced by "H-space".

---

* = Abstract of lecture given by Roy Douglas at the H-Spaces Conference in Neuchâtel, Switzerland, August, 1970.

A note with this title will appear in Topology.

University of Washington, Seattle, Washington, 98105, U. S. A.
University of British Columbia, Vancouver, B. C. Canada, and
Forschungsinstitut für Mathematik, ETH, 8006, Zürich.

INFINITE LOOP SPACES[1] - AN HISTORICAL SURVEY

James D. Stasheff[2]
Temple University

Loop spaces have long played an important role in algebraic topology. Two major themes have been the problem of computing $H_*(\Omega X)$ and the problem of recognizing a space of the homotopy type of a loop space. Historically the development has gone something like this:

In special cases $H_*(\Omega X)$ could be computed by Morse theory or by Serre's spectral sequence for the fibration $\Omega X \to PX \to X$. By (1955, 14), James had produced his reduced product which computed $H_*(\Omega SY)$. It was soon followed (1956, 2) by the Adams-Hilton machinery for computing $H_*(\Omega X)$ of a simply connected complex X, and then (1956, 1) by the Adams cobar construction for $C_*(\Omega X)$ for any simply connected X. About the same time (2), Moore was able to obtain extensive results on $H_*(\Omega^2 S^{2n+1}; Z_p)$ using constructions of Cartan.

Meanwhile (1956, 15) Kudo and Araki had initiated their study of highly iterated loop spaces $\Omega^n X$ with their description of H_n-spaces and homology operations mod 2. (This work was later extended to odd primes by Browder (1960, 10) and Dyer and Lashof (1952, 12)).

At this stage, certain dual structures had become evident, though not much use was made of the duality. On the one hand, the Kudo-Araki - Dyer-Lashof approach underlined the fact that n-fold loop spaces ΩX admitted families of maps

$$\text{(I)} \quad (W\Sigma_n)^{(r-1)} \times_{\Sigma_n} (\Omega^r X)^n \to \Omega^r X$$

---

1 Greatly revised in light of developments during and since the conference

2 Alfred P. Sloan Fellow

where Σ_n is the symmetric group and $(W\Sigma_n)^{(r-1)}$ is the (r-1)-skeleton of a Σ_n-free acyclic complex. (Actually Dyer and Lashof used the iterated join $\Sigma_n * \ldots * \Sigma_n$ instead of $(W\Sigma_n)^{(r-1)}$). Not only were such maps analogous to the maps

(II) $$(W\Sigma_n) \otimes_{\Sigma_n} C_*X \to (C_*X)^n$$

which appear in constructing the Steenrod operations, but it was clear that these latter maps were ingredients in iterating the cobar construction.

The other essential ingredient was the characterization of loop spaces provided by the work of Dold and Lashof (1958, 11) Sugawara (1957, 29), Adams and myself (1961,28).

A connected CW-space Z was shown to have the homotopy type of a loop space if and only iff the space Z admitted a compatible family of maps

(III) $$K_n \times Z^n \to Z$$

where the K_n were appropriate complexes, contractible but not acted on by Σ^n. This characterization is important because it is homotopy invariant, while admitting an associative multiplication is not.

The above listing of the various families of maps is largely an act of hindsight. These variations on the basic theme continued to interweave and reinforce each other without working directly in concert.

The attempt to iterate the cobar construction focused attention on families of higher homotopies. At Oxford in the late 50's, the emphasis was on constructing homotopies for spaces of particular interest, for example the homotopy commutativity of O(n) in O(2n) or of F(n) in F(2n), while Adams and MacLane developed a machinery for handling whole conjeries of higher homotopies in less explicit terms. The latter became the theory of PROPs and Pacts (1963, 16).

A PROP P is a category with objects 0,1,2,... satisfying the following conditions:

(a) there is given an associative bifunctor $\oplus : P \times P \to P$ such that on objects $j \oplus k = j + k$

(b) there is given an inclusion of the symmetric group Σ_K in $(P(k,k)$

(c) for $\sigma \in \Sigma_j$, $\tau \in \Sigma_K$, the morphism $\sigma \oplus \tau \in \Sigma_{j+k} \subset P(j+k, j+k)$ and is that permutation which acts as σ on the first j elements, as τ on the last k

(d) for $P_i \in P(j_i, K_i)$ and $\sigma \in \Sigma_m$,

we have

$$\sigma(K_1,\ldots,K_m)(p_1 + \ldots + p_m)$$

$$= (p_{\sigma^{-1}(1)} + \ldots p_{\sigma^{-1}(m)})\sigma(j_1,\ldots,j_m)$$

where $\sigma(K_1,\ldots,K_m) \in \Sigma_K, K = \Sigma K_i$ and acts by permuting the blocks in the given partition of K elements.

Example: End X is the PROP given by End $X(j,k) = \{X^i \to X^k\}$ with all the obvious structure.

If we have a morphism of PROPs $P \to \text{End } X$, we say that P acts on X or X is a P-space. The corresponding notion of a PACT uses differential graded Λ-modules instead of topological spaces.

In MacLane's seminar in 1967, the program had arrived at the following conjecture:

There is a PACT $\boldsymbol{\Omega}$ such that $\boldsymbol{\Omega}$ acts naturally on $C_*(X)$ so as to induce an action on the cobar construction $\Omega C_*(X)$ compatible at least up to homotopy with the natural action of $\boldsymbol{\Omega}$ on $C_*(\Omega X)$ and the usual equivalence $\Omega C_*(X) \to C_*(\Omega X)$.

A specific candidate for $\boldsymbol{\Omega}$, called the Steenrod PACT and including the actions (II) was proposed.

Meanwhile, Husseini (1962, 13) took up the idea of James' reduced product Y_∞ as the free associative H-space generated by Y and generalized the construction to obtain associative H-spaces module relations, for example.

Regarding the reduced product as a cellular model for ΩSY, Milgram (1966, 21) constructed models for $\Omega^n S^n Y$ and Dyer and Lashof (in the preprint of (12) had a model for $\Omega^\infty S^\infty Y$. The Dyer and Lashof model TY was essentially the union

$$\bigcup W\Sigma_n \times_{\Sigma_n} Y^n \ / \sim$$

under appropriate identifications as $n \to \infty$.

Work on what May now calls the "recognition principal" proceeded more slowly. The program being pursued was the following:

A commutative associative H-space was known to have the homotopy type of a product of $K(\pi,n)$'s and hence of an infinite loop space.

A loop space ΩX was shown by Sugawara (1960, 31) to be the loops on an H-space if there were suitably compatible maps

$$I^{n-1} \times (\Omega X)^n \to \Omega X$$

The hope was to construct a still larger family guaranteeing the existence of

$$K_n \times X^n \to X$$

as in (III). In MacLane's PACT seminar (1967) I gave a preliminary definition of a TOPAC or topological PACT and suggested there was a topological analogue of the (dual to the) Steenrod PACT leading to the following conjecture:

There is a TOPAC $\mathbb{B}$ such that if $\mathbb{B}$ acts on A (connected) then A has the homotopy type of ΩX and $\mathbb{B}$ acts on X, therefore X is an infinite loop space.

The following year (1968, 9) Boardman and Vogt announced success with a variation on this approach. First Boardman and Vogt define a "category of operators" which is just a topological PROP i.e. a PROP such that

e) the morphisms $P(j,k)$ constitute a topological space such that composition and $\oplus$ are continuous.

The PROP End X defined in terms of maps $X^j \to X^k$ is a topological PROP, at least in the compactly generated compact-open topology. Henceforth all PROP's will be topological.

If P acts on X, we mean P→End X to be continuous.

The following special PROPs are of importance:

The PROP $\mathcal{Q}$ describing an associative monoid: $\mathcal{Q}(n,1) = \Sigma_n$.

The PROP $W\mathcal{Q}$ with $W\mathcal{Q}(n,1) = \Sigma_n \times K_n$ where K_n is as in (III), the complex relevant to my A_n-spaces.

The PROP $W\Sigma$ with $W\Sigma(n,1) = W\Sigma_n$.

The Boardman-Vogt program can be described as follows(8,9):

They generalize the W-construction so as to 1) apply to PROP's

2) preserve homotopy type

3) have the property:

For any PROP P, WP acts on X iff WP acts on any space of the homotopy type of X.

Their work also includes tensor product of PROP's. For special PROP's they use ordinary homotopy theory to define an equivariant map $W(\mathcal{Q} \boxtimes P) \to P$ or, equally well, can define $W(\mathcal{Q} \boxtimes W\Sigma) \to P$.

Next, there is a construction (due independently to Frank Adams) which given a $W\mathcal{Q}$ action on X produces a space $M_{\mathcal{Q}}X \simeq X$ with an associative multiplication. Vogt generalizes this construction so that given an action of $W(\mathcal{Q} \boxtimes P)$ on X he constructs $M_{\mathcal{Q} \times P}X$ and an action of $\mathcal{Q} \boxtimes P$ on $Y = M_{\mathcal{Q} \boxtimes P}X$. In particular, this means Y is a monoid and P acts via $\mathrm{Hom}(Y^n, Y)$ and hence on B_Y. Now they can iterate.

Boardman chooses P judiciously in terms of classical vector space structure to get P for X=O, Top, F and their classifying spaces (9). The switch $W(\mathcal{Q} \boxtimes W\Sigma) \to P$ allows the argument to be in terms of $W\Sigma$, if so desired, after the first stage.

Meanwhile for some of these spaces Z of classical interest, Tsuchiya (1968, 31) had constructed fairly explicit maps

$$W\Sigma_n \times_{\Sigma_n} Z^n \to Z$$

and paid some attention to their compatibility. He, Milgram (23) and May (19) were able to make great strides with the computation of $H^*(BSF;Z_p)$. May (1968, 18) also cleared up certain details in the definition of "infinite loop space" in comparison to "Ω-spectra".

About the time of Boardman and Vogt's work, Beck (1968, 7) observed that $\Omega^n S^n Y$ exhibits structure familiar in another part of category theory. The canonical retraction $S\Omega Z \to Z$ by $(t,\lambda) \to \lambda(t)$ induces a retraction $\Omega^n S^n \Omega^n S^n Y \xrightarrow{\varepsilon} \Omega^n S^n Y$.

Let QY denote $\Omega^\infty S^\infty Y$; we also have $QQY \xrightarrow{\varepsilon} QY$ with the property that $\varepsilon(\varepsilon Q) = \varepsilon(Q\varepsilon)$ i.e.

$$\begin{array}{ccc} QQQ & \xrightarrow{\varepsilon Q} & QQ \\ {\scriptstyle Q\varepsilon}\downarrow & & \downarrow{\scriptstyle \varepsilon} \\ QQ & \xrightarrow[\varepsilon]{} & Q \end{array}$$

and similarly at the finite stages. In categorical language we have a triple.

Beck proves: Given a retraction $r: QX \to X$ such that $r\varepsilon = r(Qr)$ (i.e. X is a Q-algebra), then X is an infinite loop space. The map ε and its particular definition in terms of $S\Omega$ are used to construct the iterated B----BX simultaneously. (If Z is an $\Omega^n S^n$-algebra, Beck constructs B---BX up to n-factors B).

Next Barratt (1969, 5) working semisimplically rediscovered the Dyer-Lashof construction TY and refined it to give a construction ΓY which is always of the homotopy type of $\Omega^\infty S^\infty Y$. The difference is that ΓY is essentially the universal group of the monoid TY or, equivalently, ΩBTY. (In the special case $Y = S^0$, Quillen observed that QS^0 has the homotopy type of $\Omega B(\cup B_{\Sigma_n})$ where the disjoint union of the B_{Σ_n}'s is given the "Whitney sum" multiplication $B_{\Sigma_p} \times B_{\Sigma_q} \to B_{\Sigma_{p+q}}$, while Priddy, Barratt and Kahn (1970, 26) showed: there is a canonical map of B_{Σ_∞} in the component of the constant loop of QS^0 inducing isomorphisms of homology (the fundamental group map is $\Sigma_\infty \to Z_2$).

Last spring (1970, 27) Segal came up with a different way of handling the homotopies involved, motivated perhaps by Quillen's analysis of $\Omega^\infty S^\infty$ in terms of $B\Sigma_n$. The motivation may also be the "classic" one that the Whitney sum is associative if we consider $R^p + R^q = R^{p+q}$, but involves all sorts of homotopies as we stabilize. (This specific problem had interfered with attempts at showing F to be an infinite loop space back in 1959; even the homotopy associativity of BF required special arguments in (24).)

Segal's method: (27, cf. Anderson's version, 3) handles the higher homotopies by not using a single space X and its powers X^n but rather a family X_i such that $X_n \simeq (X_1)^n$. A direct comparison of the methods should be possible in terms of a generalized notion of PROP capable of handling such families of spaces. (Compare Boardman's colored trees (8).)

In a still more recent preprint of Anderson, the essence of the approach appears more accessibly to be:

(1) the use of categories $\mathcal{C}$ as a generalization of monoid (a monoid can be regarded as a category with one object) with the corresponding generalization of the "classifying space" $B\mathcal{C}$

(2) the identification of natural transformations of functors with homotopies of maps of "classifying spaces"

(3) the introduction of a tensor product category $X \boxtimes \mathcal{C}$ for any set X. For certain categories $\mathcal{C}$ called permutative, this is functorial in X and hence can be defined for a simplicial set X. If $\mathcal{C} = \{\mathcal{C}(n)\}$ is a simplicial category, then $X \boxtimes \mathcal{C}$ can be regarded as a bisimplicial category.

(4) the proof that diag $G\bar{W}\bar{B} \simeq \Omega^n$ diag $G\bar{W}\bar{B}(S^n \boxtimes \mathcal{C})$ using Quillen's spectral sequence which gives the homotopy groups of the diagonal of a bisimplicial complex in terms of the horizontal homotopy groups of the vertical homotopy groups.

The permutative categories $\mathcal{C}$ are those with a functor $+: \mathcal{C} \times \mathcal{C} \to \mathcal{C}$ which has a neutral object, is associative and for $\sigma \in \Sigma_n$, $A_i \in \mathcal{C}$, $A_1 + \dots + A_n$ is naturally equivalent to $A_{\sigma(1)} + \dots + A_{\sigma(n)}$ the equivalence being uniquely determined by a choice for n = 2 and any decomposition of σ into transpositions. Examples occur whenever + is truly direct product.

In addition to giving an alternate approach to recognizing infinite loop spaces, it was Segal's presentation which called attention to the fact that a strong enough Dyer-Lashof structure could exist only in an infinite loop space: X is an infinite loop space iff there exists a suitably compatible family of maps of form (I):

$$\bigcup W\Sigma_n \times_{\Sigma_n} X^n \to X.$$

Actually Segal states the above theorem in a form shown to be equivalent by Quillen: A representable homotopy functor is a cohomology theory iff the functor admits a trace (or transfer or Gysin map). A trace τ for a representable homotopy functor D to abelian groups is a functor defined on coverings of finite index. Since D is a homotopy functor, axiomatic properties of τ provide homotopies between representing maps. Roush (1970 - Princeton thesis) has shown sufficiently strong axioms on the trace for a cohomology theory imply the theory is a product of ordinary ones, i.e. the theory is represented by a product of $K(\pi,n)$'s.

As shown by Quillen, a minimal property of trace corresponds to a family of form (I) for the representing space. The first degree of compatibility desired is with respect to inclusion of n-factors in n+1 - at least up to homotopy. This corresponds to a retraction $TZ \to Z$, and this at least we can verify for the maps constructed by Tsuchiya (31), Milgram (22), or Madsen (17).

The most we could demand would be that this retraction $TZ \to Z$ make Z a T-algebra. This requires compatibility precisely so as to make Z a $W\Sigma$-space. Such strict compatibility is of course more than one could expect from general homotopy nonsense; some more rigid "geometry" would be needed.

What one could hope to construct up to homotopy is a family (I) giving rise to a retraction $TZ \to Z$ making Z a "strongly homotopy T-algebra". Perhaps the lack of precise compatibility in the maps constructed by Tsuchiya or Milgram or Madsen can be handled this way. The work of Vogt on homotopy P-maps may provide the machinery.

Doing constructions with such generality is however, not a pleasant prospect; May now (1970, 20) offers a very pleasant compromise. He uses some of the best of both worlds, PROP's and triples. He shows that a (standard) PROP determines a triple (perhaps the converse is true) and thus is able to relate the geometrically constructed Boardman PROP's very neatly to the triple $\Omega^n S^n$ and hence to generalize Beck's construction. More precisely, he shows the triple C_n associated to Boardman's "little cubes" category $(C_n(9)$ maps by a map of triples into $\Omega^n S^n$ preserving homotopy type. Thus he shows $C_n Y \simeq \Omega^n S^n Y$ for $n \leq \infty$, giving new and useful models for $\Omega^n S^n Y$. He uses these equivalences to construct an n-fold classifying space $B^n X$ for a C_n-space X, by a generalization of Beck's construction. (The method is <u>not</u> iterative).

He then uses the direct product of PROP's to relate any X with any E_∞-PROP action to this construction and hence construct $B^n X$ (although C_n-action is not a homotopy invariant, it is very useful computationally.

Beck has now (6) another approach. For any E_∞-PROP E, he defines a suspension within the category of E-spaces. (essentially as a quotient of the free E-space generated by the ordinary suspension). He then shows that this gives a classifying space for X. The iterated classifying space is then the iterated E-suspension! (The relation to the Segal-Anderson classifying space obtained by condensation deserves attention. The definition of the quotient of the free E-space is reminiscent of Husseini's constructions of reduced product type being "free modulo relations").

At the present time, the main consequence of knowing that F or submonoids or quotients have iterated classifying spaces is the availability of Dyer Lashof operations for computing homology or cohomology. The Tsuchiya approach is to construct the corresponding maps directly, but to prove appropriated "distributivity" with respect to the QS^0 structure rather than to attempt to verify a WΣ-action is obtained. On the other hand, May points out that in order to compute such operations one does not need maps precisely of the form (I) but rather can replace $W\Sigma_n$ by any contractible Σ_n-free space, e.g. E(n,1) in any E_∞-PROP. Moreover, with May's technique there appears naturally a map of QS^0 into the infinite loop space with which the operations are automatically compatible.

In fact in most of the above approaches we construct such maps, the data for the operations, before constructing the iterated classifying spaces. In turn, we have learned how to compute Dyer-Lashof operations and then to recognize the infinite loop space structure they reflect. Perhaps now we can look for applications (beyond the existence of Dyer-Lashof operations) for the iterated classifying spaces we now know to exist for these spaces which are geometrically so significant.

REFERENCES

[1] Adams, J. F., On the cobar construction, Colloque de Topologie Algebrique, Louvain (1956) 81-87

[2] Adams, J. F., Hilton, P. J., On the chain algebra of a loop space, Comm. Math. Helv. 30 (1956) 305-330.

[3] Anderson, D. W., Spectra and Γ-sets, AMS conference on Algebraic Topology, Madison (1970)

[4] Anderson, D. C., ICM, Nice, 1970.

[5] Barratt, M. G., A free group functor for stable homotopy, AMS Conference on Algebraic Topology, Madison (1970)

[6] Beck, J., Classifying spaces for homotopy-everything H-spaces, (preprint - hopefully to appear in the Proceedings of this Conference).

[7] Beck, J., On H-spaces and infinite loop spaces, Category Theory, Homology Theory and their Applications III, Springer-Verlag (1969) 139-153

[8] Boardman, J. M., Homotopy structures and the language of trees, AMS Conference on Algebraic Topology, Madison (1970).

[9] Boardman, J. M., and Vogt, R. M., Homotopy-everything H-spaces, Bull. AMS 74 (1968), 1117-1122.

[10] Browder, W., Homotopy commutative H-spaces, Annals of Math(2) 75 (1962), 283-311.

[11] Dold, A., Lashof, R., Principal quasifibrations and fibre homotopy equivalences of bundles, Ill. J.M. 3 (1959), 285-305.

[12] Dyer, E., Lashof, R., Homology of iterated loop spaces, AJM 84 (1962), 35-88

[13] Husseini, S., Constructions of the reduced product type. Topology 2(1963), 213-237, II. Topology 3(1965), 59-79.

[14] James, I., Reduced product spaces, Annals of Math(2) 62 (1955), 170-197.

[15] Kudo, T., Araki, S., Topology of H_n-spaces and H-squaring operations, Mem. Fac. Sci. Kyushu U. Ser. A. 10 (1956)85-120.

[16] MacLane, S., Categorical Algebra, Chapter IV, Colloquium Lectures, AMS, 1963, 26-29.

[17] Madsen, I., On the action of the Dyer-Lashof algebra in $H_*(G)$ and $H_*(G/Top)$, Thesis, Univ. of Chicago (1970)

[18] May, J., Categories of spectra and infinite loop spaces, Category Theory, Homology Theory and their applications, III, Lecture Notes in Mathematics 99, Springer-Verlag, Berlin (1969) 448-479.

[19] May, J., Homology operations in loop spaces, AMS, Conference on Algebraic Topology, Madison (1970)

[20] May, J. P., The geometry of iterated loop spaces, preprint.

[21] Milgram, R. J., Iterated loop spaces, Ann. of Math. 84 (1968), 386-403.

[22] Milgram, R. J., Symmetries and operations in homotopy theory, AMS Conference on Algebraic Topology, Madison (1970)

[23] Milgram, R. J., The mod two spherical characteristic classes, Annals of Math. 92 (1970) 238-261,, c.f. The cohomology of B_G, Proc. conf. on Alg. Top., Chicago Circle (1968) 213-220.

[24] Milnor, J., On characteristic classes for spherical fibrations, Comm. Math. Helv. 43 (1968), 51-77.

[25] Moore, J. C., The double suspension and p-primary components of the homotopy groups of spheres, Bol. Soc. Mat. Mex.1(1956) 28-37.

[26] Priddy, S., On $\Omega^\infty S^\infty$ and the infinite symmetric join, AMS conference on Algebraic Topology, Madison (1970).

[27] Segal, G., Homotopy-everything H-spaces (preprint).

[28] Stasheff, J., Thesis, Oxford University, 1961.

[29] Sugawara, M., A condition that a space is group like. M. J. Okayama U. 7 (1957), 123-149.

[30] Sugawara, M., On the homotopy-commutativity of groups and loop spaces. Mem. Coll. Sci. Univ. Kyoto Ser. A. Math. 33 (1960.61), 257-269.

[31] Tsuchiya, A., Characteristic classes for spherical fiber spaces, Proc. Jap. Acad. 44 (1968), 617-622 and preprint.

CLASSIFYING SPACES FOR HOMOTOPY-EVERYTHING H-SPACES

Jon Beck
Cornell and Aarhus Universities

I show that homotopy-everything H-spaces (Boardman-Vogt [2]) can be de-looped in the following simple way. The category of homotopy-everything spaces, or E-spaces, is closed under the ordinary loop space functor. There exists a "suspension" operation on E-spaces, denoted by $S^1 \otimes X$. This intrinsic "suspension" on E-spaces is left adjoint to the loop space functor. Then the adjointness map

$$X \to \Omega(S^1 \otimes X)$$

is a weak homotopy equivalence for every E-space X which has $\pi_0(X)$ a group. Thus the classifying space of an E-space is nothing but its "suspension" in the category of E-spaces.

The suspension is a special case of a more general tensor product $A \otimes X$ defined between a topological space A and E-space X. It is shown that $h_n(A) = \pi_n(A \otimes X)$ is a generalized homology theory. Both this and the de-looping equivalence follow from the fact that the tensor product converts cofibrations into fibrations (2.4). This last is proved by a method of Dold-Thom [3].

I am grateful to Michael Barr and Peter May for helpful comments.

1. Examples of suspensions in topological categories. The main tool is the concept of a topological category, $\mathfrak{X}$. By this is meant [1] that $\mathfrak{X}$ is a category which has functorial operations

$$\mathfrak{X}^{op} \times \mathfrak{X} \xrightarrow{X,\ Y \to Y^X} \underline{Top},$$

$$\underline{Top}^{op} \times \mathfrak{X} \xrightarrow{A,\ Y \to Y^A} \mathfrak{X},$$

$$\underline{Top} \times \mathfrak{X} \xrightarrow{A, X \to A \otimes X} \mathfrak{X}$$

such that adjointness 1-1 correspondences hold between mappings

$$A \xrightarrow{f} Y^X \text{ in } \underline{Top},$$

$$X \xrightarrow{f'} Y^A \text{ in } \mathfrak{X},$$

$$A \otimes X \xrightarrow{f''} Y \text{ in } \mathfrak{X}.$$

In particular if $A = S^1 \in \underline{Top}$ is the circle, we have the suspension $S^1 \otimes X$ and the loop object $\Omega X = X^{S^1}$ of an object $X \in \mathfrak{X}$. These are again members of $\mathfrak{X}$. By adjointness there is a natural

$\mathcal{X}$-morphism

$$X \longrightarrow \Omega(S^1 \otimes X).$$

Top, the category of topological spaces with base point, is a topological category. Y^X, Y^A are just the usual function spaces (X, Y, A are all topological spaces), and the smash product $A \wedge X = A \times X/(A \vee X = 0)$ plays the role of the tensor product.

Topological groups form another example of a topological category. If G, H are groups there is the space of continuous homomorphisms H^G. This is a Top-valued hom functor so we are encouraged to seek adjointness relations among maps

$$A \xrightarrow{\ f\ } H^G \quad \text{in Top},$$

$$G \xrightarrow{\ f'\ } H^A \quad \text{in Top Gps},$$

$$A \otimes G \xrightarrow{\ f''\ } H \quad \text{in Top Gps}.$$

H^A is the group of all continuous maps $A \to H$ with pointwise group operations. The less well known tensor product $A \otimes G \in$ Top Gps is easy to construct. It is the free topological group generated by the smash product $A \wedge G \in$ Top modulo "tensor relations" $(a \wedge g)(a \wedge g') = a \wedge gg'$.

In the semisimplical case this tensor product has long been known [5]. In the topological case the construction is eased by employing some nice category such as compactly-generated spaces. Then the free topological group is just the ordinary free group furnished with an appropriate topology. This topology can be learned from Dold-Thom [3].

If we specialize to $A = S^1 \in$ Top, we get a 1-1 correspondence between topological group homomorphisms

$$G \xrightarrow{\ f\ } H^{S^1} = \Omega H,$$

$$S^1 \otimes G \xrightarrow{\ g\ } H.$$

Because of this adjointness we can certainly refer to $S^1 \otimes G$ as the suspension of G within the category of groups. Note that ΩH has its group structure induced by that of H, but this is homotopic to the usual end-to-end composition of loops. By adjointness we have the canonical topological group homomorphism

$$G \xrightarrow{\ \eta\ } \Omega(S^1 \otimes G).$$

Explicitly this map is $\eta(g)(t) = t\wedge g$ where $t \in S^1$ and $t\wedge g$ is an element of the smash product which sits inside of the tensor product.

Example. Let $G = \mathbf{Z}$, the discrete group of integers. Note that $\mathbf{Z} = F(S^0)$ is the free topological group generated by the pointed space S^0 (zero-sphere). This implies $S^1 \otimes \mathbf{Z} = F(S^1)$, the free topological group on S^1.

(Proof: topological group homomorphisms $S^1 \otimes \mathbf{Z} \to H$ are adjoint to continuous maps $S^1 \to H^{\mathbf{Z}} = H$, and these are adjoint to homomorphisms $F(S^1) \to H$. Since an object in a category is uniquely determined up to isomorphism by its maps into all other objects, $S^1 \otimes \mathbf{Z} = F(S^1)$.)

Now, up to homotopy, $F(S^1) \simeq \Omega\Sigma(S^1)$ (proof in [4, 1]). In fact it is easy to check that

$$\mathbf{Z} \longrightarrow \Omega(S^1 \otimes \mathbf{Z})$$

is a map $\mathbf{Z} \to \Omega^2 S^2$ which associates to $n \in \mathbf{Z}$ a map $S^2 \to S^2$ of degree n.

In general, η should not be a homotopy equivalence because $\Omega(S^1 \otimes G)$ as the loop space of a group should have a higher degree of homotopy commutativity than G. Unless, of course, G is infinitely homotopy commutative to start with, which leads to the grouplike homotopy-everything spaces discussed in the next section.

2. Suspensions of homotpy-everything spaces. Let E be one of the Boardman-Vogt categories of operators defining homotopy-everything H-spaces [2]. Recall that $E(n,1)$, the space of additions of n distinct variables, is contractible and admits an action of the symmetric group S_n which can be thought of as permuting the variables. This implies homotopy commutativity, for example, for if $\theta \in E(2,1)$ is an addition $x + y$, then the transposition $\tau(\theta)$ is the addition $y + x$, and by contractibility these two operations are homotopic. Higher homotopy commutativities are found by considering operation spaces $E(n,1)$ for larger values of n. Higher homotopy associativities also result from this definition.

X is an E-space if the operations in $E(n,1)$ are continuously represented as maps $X^n \to X$. This makes X into an H-space equipped with homotopy commutativities and associativities. $f : X \to Y$ is an E-homomorphism if all diagrams

$$\begin{array}{ccc} X^n & \xrightarrow{f^n} & Y^n \\ \theta\downarrow & & \downarrow\theta \\ X & \xrightarrow{f} & Y \end{array}$$

strictly commute. This category of E-spaces is denoted by $\underline{Top}^E$.

(2.1) Theorem. $\underline{Top}^E$ is a topological category.

Indeed, if X, Y are E-spaces, take Y^X as the topological space of all E-homomorphisms $X \to Y$. If $A \in \underline{Top}$, $Y \in \underline{Top}^E$, the space of all continuous maps Y^A is an E-space under the pointwise E-operations, i.e. if $f_1, \ldots, f_n \in Y^A$, $\theta \in E(n,1)$, then $\theta(f_1, \ldots, f_n) \in Y^A$ is the composite

$$A \xrightarrow{\Delta} A_n \xrightarrow{f_1 x \ldots x f_n} Y^n \xrightarrow{\theta} Y .$$

$A \otimes X$ is the free E-space generated by $A \wedge X$ modulo tensor relations $\theta(a \wedge x_1, \ldots, a \wedge x_n) = a \wedge \theta(x_1, \ldots, x_n)$. As to the existence of free E-spaces $F(A)$ where $A \in \underline{Top}$, that follows from category theory. Suffice it to say that the elements of $F(A)$ can be written in the form $\theta(a_1, \ldots, a_n)$ and topologized by an identification map (again compare [3])

$$\bigcup_{n \geq 0} E(n,1) \times A^n \to F(A).$$

Note that the above $a_1, \ldots, a_n$ need not be distinct elements of A. The repetitions give rise to the homotopical non-triviality of homotopy-everything spaces.

One easily verifies that adjointness holds between maps $X \to Y^A$ and maps $A \otimes X \to Y$ in $\underline{Top}^E$.

The suspension of $X \in \underline{Top}^E$ is $S^1 \otimes X$, and the loop space of X is $\Omega X = X^{S^1}$. By adjointness we have the canonical E-homomorphism

$$X \longrightarrow \Omega(S^1 \otimes X).$$

(2.2) Delooping Theorem. $X \to \Omega(S^1 \otimes X)$ is a weak homotopy equivalence for any E-space X such that $\pi_0(X)$ is a group (call these grouplike E-spaces).

You can de-loop n times at once by using the adjointness homotopy equivalence $X \to \Omega^n(S^n \otimes X)$. The inductive proof of this goes:

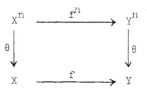

$$\begin{array}{ccc} X & \xrightarrow{\eta_n} \Omega^n(S^n \otimes X) \xrightarrow{\Omega^n\eta} & \Omega^n\Omega(S^1 \otimes S^n \otimes X) \\ & \searrow^{\eta_{n+1}} & \downarrow = \\ & & \Omega^{n+1}(S^{n+1} \otimes X) \end{array}$$

Thus $(S^n \otimes X)_{n\geq 0}$ is an Ω-spectrum. Up to homotopy any Ω-spectrum arises from an E-space in this way, and any connective cohomology theory on CW-complexes [2,6].

There is a more direct relation between homology theories and E-spaces:

(2.3) Theorem. If X is a grouplike E-space, then $h_n(A) = \pi_n(A \otimes X)$ is a generalized homology theory on CW-complexes. Any connective homology theory arises in this way.

The proofs of (2.2) and (2.3) are based on the fact that the tensor product operation $(\ ) \otimes X$ essentially converts cofibrations into fibrations (see (2.4)). We tensor the cofibration $S^0 \to I \to S^1$ with X and obtain the following diagram:

$$\begin{array}{ccc} X = S^0 \otimes X & \xrightarrow{\eta} & \Omega(S^1 \otimes X) \\ \downarrow & & \downarrow \\ I \otimes X & \xrightarrow{\eta_1} & E(S^1 \otimes X) \\ \downarrow & & \downarrow \\ S^1 \otimes X & \xrightarrow{=} & S^1 \otimes X. \end{array}$$

E above is the contractible path space functor and $\eta_1(t\wedge x)(u) = tu\wedge x$. The map induced on the fibers is the canonical η: $X \to \Omega(S^1 \otimes X)$. Since $I \otimes X$ is contractible, it follows from the homotopy exact sequence that η is a weak homotopy equivalence.

Thus (2.2) and (2.3) result from the following

(2.4) Fibration Theorem. Let $A \to B \to B/A$ be sub and quotient complex. If A is connected or $\pi_0(X)$ is a group, then $A \otimes X \to B \otimes X \to (B/A) \otimes X$ is weakly homotopy equivalent to a fibration.

Here is an outline of the proof. As suggested by [8, Lemma 7] we first deform the E-structure of X so that

$$\theta(o,\ldots,o,x,o,\ldots,o) = x$$

holds for all $\theta \in E(n,1)$ and vectors with at most one component $x \neq o$ (base point). This is called normalizing X.

From now on we assume X is normalized. Note that this does not affect the homotopy types of the tensor products $A \otimes X$, etc.

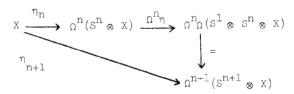

Second, we normalize the category of operators E. This is done by a morphism of categories of operators $E \to E'$ which is induced by contracting $E(1,1)$ to a single point. E' is still a homotopy-everything category of operators. Let $F(A), F'(A)$ be the free E, E'-spaces generated by a space A. If A is a CW complex we have:

(2.5) The natural map $F(A) \to F'(A)$ is a weak homotopy equivalence.

Since X has been normalized there is a unique E'-structure on X such that the normalized E-structure is induced by the map $E \to E'$. For every A there results a natural map of our previous tensor product $A \otimes_E X \to A \otimes_{E'} X$. With obvious notation, (2.5) implies:

(2.6) $A \otimes X \to A \otimes' X$ is a weak homotopy equivalence.

Finally, using naturality, we reduce (2.4) to a known context:

$$\begin{array}{ccccc} A \otimes X & \to & B \otimes X & \to & (B/A) \otimes X \\ \simeq\downarrow & & \simeq\downarrow & & \simeq\downarrow \\ A \otimes' X & \to & B \otimes' X & \to & (B/A) \otimes' X. \end{array}$$

(2.7) The bottom sequence of $\otimes'$-products is a quasifibration.

(2.7) is proved classically, like [3, 5.2]. First revert to the notation E for the normalized homotopy-everything category of operators and $A \otimes X$ for the tensor product. Filter $A \otimes X$ by letting F_n be the subspace consisting of all elements that can be written in the form $\theta(a_1 \wedge x_1, \dots, a_n \wedge x_n)$ for some $\theta \in E(n,1)$. Inductively assume that p_n is a quasifibration

$$\begin{array}{ccc} p^{-1}F_n((B/A) \otimes X) & \to & p^{-1}F_{n+1}((B/A) \otimes X) \\ p_n \downarrow & & p_{n+1} \downarrow \\ F_n((B/A) \otimes X) & \to & F_{n+1}((B/A) \otimes X) \end{array}$$

where $p\colon B \otimes X \to (B/A) \otimes X$.

The subcomplex $A \subset B$ is a strong deformation retract of an open neighborhood u. Let $U \subset F_{n+1}((B/A) \otimes X)$ be the open set consisting of all $\theta_{n+1}(b_1 \wedge x_1, \dots, b_{n+1} \wedge x_{n+1})$ with at least one

$b_i \in u/A \subset B/A$. U can be deformation-retracted onto $F_n((B/A) \otimes X)$. There is a covering deformation in $p^{-1}(U)$. Its effect on a fiber can be identified with translation by an element $\alpha \in A \otimes X$.

If A is connected, so is $A \otimes X$. If $\pi_0(X)$ is a group, so is $\pi_0(A \otimes X)$. In both cases translation is a homotopy equivalence. Thus $p|p^{-1}(U)$ is a quasifibration [3, 2.10].

Here translation by α is the map $y \to \theta_2(\alpha, y)$ where θ_2 is some fixed "addition" in $E(2,1)$. We use θ_2 to identify the rest of our map with a product projection.

Let $V \subset F_{n+1}((B/A) \otimes X)$ be the open set $F_{n+1} - F_n$. There is a well-defined map

$$\begin{array}{ccc} p^{-1}(V) & \rightarrow & (A \otimes X) \times V \\ & {\scriptstyle p} \searrow \quad \swarrow {\scriptstyle \text{proj.}} & \\ & V & \end{array}$$

which sends $\theta(a_1 \wedge x_1, \ldots, b_m \wedge x_m, \ldots)$ into $A \otimes X$ or V by setting all $a = o$ or $b = o$. Its inverse is $(\alpha, v) \to \theta_2(\alpha, v)$. Using the homotopy-everything property, one verifies that this is a fiber homotopy equivalence (in contrast with [3] where it is a homeomorphism).

Thus p_{n+1}, being a quasifibration over U, V and $U \cap V$, is a quasifibration. Since $p = \varinjlim p_n$, p is also a quasifibration.

The point of normalization is that otherwise the above maps and homotopies meander from fiber to fiber in a disconcerting fashion. The theorem can be proved without normalizing but then a rather depressing covering homotopy property must be invoked. An axiomatic approach is really needed.

In case $E(1,1)$ is a single point, normalization is also unnecessary, for example $E = WS_*$ where $WS_*(n,1)$ is the contractible bar construction WS_n on the symmetric group.

3. Other Examples. For topological groups (2.4) is false as already noted. However, the sequence $G \to I \otimes G \to I \otimes G/G$ gives a classifying space for G (announced in [1] and proved like [3, 5.4]. Note that this differs from the many other classifying space constructions in that the contractible total space is also a group. The same approach should work for monoids or A_∞-spaces.

In the case of commutative monoids the $\otimes$ product maps

$$\underline{\text{Top}} \times \underline{\text{Top}}^{SP} \to \underline{\text{Top}}^{SP}$$

the category of operators appropriate here "being" the symmetric product. $A \otimes X \to B \otimes X \to (B/A) \otimes X$ a quasifibration slightly generalizes [3, 5.2] where $X = SP(S^0)$ and $A \otimes X = A \otimes SP(S^0)$ $= SP(A \wedge S^0) = SP(A)$. The same result, with principal bundles, holds for commutative groups. Our $A \otimes X$ is then $B(X,A)$ of [6].

The homotopy-everything category best related to Ω-spectra is that of Q-spaces [1]. Theorem (2.4) is true for Q-spaces but the proof is more elaborate since Q-spaces do not come from a finitary category of operators.

References

[1] J. Beck, On H-spaces and infinite loop spaces. Category Theory, Homology Theory and their Applications, III (Batelle Conference), Lecture Notes in Math. 99 (1969), Springer Verlag, 139-153.

[2] J. M. Boardman - R. Vogt, Homotopy-everything H-spaces, Bull. AMS 74 (1968), 117-1122.

[3] A. Dold - R. Thom, Quasifaserungen und unendliche symmetrische Produkte, Ann. of Math. 67 (1958), 239-281.

[4] I. M. James, Reduced product spaces, Ann. of Math 62 (1955), 170-197.

[5] D. M. Kan, On homotopy theory and c.s.s. groups, Ann. of Math. 68 (1958), 38-53.

[6] J. P. May, Categories of spectra and infinite loop spaces, ibid. as [1], 448-479.

[7] M. C. McCord, Classifying spaces and infinite symmetric products, Transactions AMS 146 (1969), 273-297.

[8] J. D. Stasheff, Homotopy associativity of H-spaces. I, Transactions AMS 108 (1963), 275-292.

THE UNIVERSAL H-SPACE OF A MONOID[1]

GERALD J. PORTER[2]

University of Pennsylvania

Let M be a monoid. For $m \in M$, define $L_m : M \longrightarrow M$ by $L_m(x) = mx$. If L_m is a homotopy equivalence for all $m \in M$ we call M an H-space. For associative H-spaces Dold and Lashof [2] have constructed a principal quasi-fibration, $M \longrightarrow EM \longrightarrow BM$, in which EM is contractible. BM is called the classifying space of M and if M has the homotopy type of a CW-complex, ΩBM is homotopy equivalent to M. Moreover this equivalence can be given by SHM maps (see below).

If M is an associative monoid which is not an H-space the Dold-Lashof construction is still possible however the resulting sequence need not be a quasi-fibration and M need not be homotopy equivalent to ΩBM.

It is the purpose of this note to show that ΩBM is universal for M with respect to SHM-maps from M to associative H-spaces (i.e. ΩBM is the universal H-space of M). We shall assume throughout that all spaces have the homotopy type of a CW-complex.

<u>Definition 1</u>: A monoid[3] is a based topological space, M, together with a continuous map, $\mu: M \times M \longrightarrow M$, such that $\mu(*,x) = \mu(x,*) = x$ for all $x \in X$ ($*$ is the base point of M). The monoid is said to be associative if $\mu(1 \times \mu) = \mu(\mu \times 1)$.

---

1 One purpose of this note is to suggest distinct uses for "monoid" and "H-space."

2 Supported in part by the National Science Foundation.

3 Monoid has been used in the literature to mean associative. We do not include associative in the definition.

Example: Let X^X be the space of continuous maps from a topological space, X, to itself under the compact open topology. The basepoint is chosen to be 1_X. Under the operation of composition, X^X is an associative monoid.

Definition 2: An (associative) H-space, X, is an (associative) monoid such that $\mu(x, \ ): X \longrightarrow X$ is a homotopy equivalence for all $x \in X$.[4]

The monoids of the example above are in general not H-spaces. However if M is path-connected it is easily seen that M is an H-space. More generally we have

Proposition 1: If X is a homotopy associative monoid then $\mu(x, \ )$ is a homotopy equivalence for all x if and only if $\pi_0(X)$ is a group.

Proof: If X is a homotopy associative H-space then it is well known that $\pi_0(X)$ is a group. We prove the converse.

Let $x \in X$ and $[x]$ be the class of x in $\pi_0(X)$. Since $\pi_0(X)$ is a group there is $y \in X$ such that $[y][x] = [*] = [x][y]$ where juxtaposition indicates the multiplication induced by μ.

Set $L_x = \mu(x, \ )$. Since $L_y L_x \sim L_{\mu(y,x)}$ we have

$$L_y L_x \sim L_{\mu(y,x)} \sim L_* = 1_X \quad \text{and}$$

$$L_x L_y \sim L_{\mu(x,y)} \sim L_* = 1_X .$$

4 Perhaps X should be called a left H-space in this case. Proposition 1 shows that if X is homotopy associative then left, right, and two-sided H-spaces are the same.

<u>Corollary</u>: In a homotopy associative monoid $\mu(x,\ )$ is a homotopy equivalence for all x if and only if $\mu(\ ,x)$ is.

<u>Definition 3</u>: An SHM-map (strongly homotopy multiplicative) $H: M \longrightarrow M'$, between associative monoids is a collection of maps $H = \{H_n : M^{n+1} \times I^n \longrightarrow M',\ n = 0,1,2,\ldots\}$ satisfying

$$H_n(m_0,\ldots,m_n,t_0,\ldots,t_n)$$
$$= \begin{cases} H_{n-1}(m_0,\ldots,m_j m_{j+1},\ldots,m_n,t_0,\ldots,\hat{t}_j,\ldots,t_{n-1}) & \text{if } t_j = 0 \\ H_j(m_0,\ldots,m_j,t_0,\ldots,t_{j-1})\cdot H_{n-j}(m_{j+1},\ldots,m_n,t_{j+1},\ldots,t_{n-1}) & \text{if } t_j = 1 \end{cases}$$

SHM-maps were originally defined by Sugawara [5] and are studied by Drachman [3] and Fuchs [4]. For most of what follows we take [4] as our reference.

<u>Definition 4</u>: Two SHM-maps H and H' are called SHM-homotopic if there is a continuous family of SHM-maps H^t, $0 \leq t \leq 1$, such that $H^0 = H$, and $H^1 = H'$.

The composite of SHM-maps may be defined ([4], p. 200) and is again an SHM-map.

<u>Proposition 2</u>: ([4], p. 203) Associative H-spaces and homotopy classes of SHM-maps form a category, $\mathcal{H}_a$.

Similarly one sees

<u>Proposition 3</u>: Associative monoids and homotopy classes of SHM-maps form a category, $\mathfrak{M}_a$.

The Dold-Lashof [2] construction assigns a 'classifying space' BX to each associative monoid X. Let $\mathcal{T}$ be the category of based topological spaces and homotopy classes of maps.

Proposition 4: The Dold-Lashof construction, B, is a functor from $\mathfrak{M}_a$ to $\mathcal{T}$.

Proof: This is proven in [4] (p. 212) for $\mathcal{H}_a$. The proof is identical for $\mathfrak{M}_a$.

Proposition 5: ([4], p. 220) $B: \mathrm{Hom}_{\mathcal{H}_a}(X,Y) \longrightarrow \mathrm{Hom}_{\mathcal{T}}(BX,BY)$ is a bijection.

Definition 5: The universal H-space of an associative monoid M, $UH(M)$, is an associative H-space together with an SHM-map $H: M \longrightarrow UH(M)$ such that if $F: M \longrightarrow X$ is an SHM-map from M to an associative H-space, X, then there is a unique (to SHM homotopy) SHM-map $G: UH(M) \longrightarrow X$ such that GH is SHM-homotopic to F.

If $UH(M)$ exists it is easily seen to be unique (up to SHM-homotopy equivalence).

Let ΩX be the Moore loop space of X.

Theorem: $\Omega BM = UH(M)$.

Proof: The adjoint of $\Sigma M \subset BM$ can be extended to an SHM-map $M \longrightarrow \Omega BM$ ([4], p. 213). Moreover this may be done so that if $F: M \longrightarrow M'$ is an SHM-map then

$$\begin{array}{ccc} M & \xrightarrow{H} & \Omega BM \\ \downarrow{\scriptstyle F} & & \downarrow{\scriptstyle \Omega BF} \\ M' & \xrightarrow{H'} & \Omega BM' \end{array}$$

commutes.

We are now in a position to prove the theorem. We wish to show that ΩBM and $H: M \longrightarrow \Omega BM$ constitute a universal H-space for M. Let $F: M \longrightarrow X$ be an SHM-map where X is an associative H-space. Since X is an H-space, $H': X \longrightarrow \Omega BX$ is a homotopy equivalence and has an SHM inverse $\overline{H}: \Omega BX \longrightarrow X$ ([4], p. 205). Set $G = \overline{H}\cdot\Omega BF$. $\overline{H}\cdot\Omega BF\cdot H = \overline{H}\cdot H'\cdot F \sim F$. Thus G has the desired property.

We must still prove that G is unique up to SHM-homotopy. Assume there is G' such that $G'\cdot H \sim G\cdot H \sim F$. Applying B we get $BG'\cdot BH \sim B(G'\cdot H) \sim B(G\cdot H) \sim BG\cdot BH$. Since BH is a homotopy equivalence this implies $BG' \sim BG$ and the result follows from Proposition 5.

<u>Corollary</u>: Let $i: \mathcal{H}_a \longrightarrow \mathfrak{M}_a$ be inclusion then ΩB is a left adjoint for i; i.e. $\mathrm{Hom}_{\mathfrak{M}_a}(X, iY) = \mathrm{Hom}_{\mathcal{H}_a}(\Omega BX, Y)$.

<u>Remark</u>: We note that if M is a discrete monoid it is not necessarily true that ΩBM will be discrete, i.e., that BM will be a $K(G,1)$ for some G (see Barratt [1]).

References

[1] M. G. Barratt, A note on the cohomology of semigroups, J. Lond. Math. Soc. 36 (1961), 496-498.

[2] A. Dold and R. Lashof, Principal quasifibrations and fibre homotopy equivalence of bundles, Ill. J. Math. 3 (1959), 285-305.

[3] B. Drachman, A generalization of the Steenrod classification theorem to H-spaces, Trans. AMS 152 (1970).

[4] M. Fuchs, Verallgemeinerte Homotopie-Homomorphismen und klassifizierde Räume, Math. Ann. 161 (1965), 197-230.

[5] M. Sugarwara, On the homotopy-commutativity of groups and loop spaces, Mem. Coll. Sci. Univ. Kyoto Ser. A Math 33 (1960), 257-269.

CATEGORICAL CONSTRUCTIONS IN ALGEBRAIC TOPOLOGY

Allan Clark
Brown University

I want to give here a brief exposition of two general categorical constructions of which many examples occur in algebraic topology. Although the connection with the theme of this conference is tenuous, both the classifying space of an associative H-space and Milgram's approximations' to iterated loop spaces |1| are particular instances of one of these constructions.

Suppose that Γ is a small category, Γ^* its dual, and $\mathcal{C}$ an arbitrary category. Then we can form the functor categories $|\Gamma,\mathcal{C}|$ and $|\Gamma^*,\mathcal{C}|$. Note that $|\Gamma,\mathcal{C}|^* = |\Gamma^*,\mathcal{C}^*|$. With reasonable restrictions on $\mathcal{C}$, I shall define two functors

$$\mathrm{Hom}_\Gamma : |\Gamma,\mathcal{C}|^* \times |\Gamma,\mathcal{C}| \longrightarrow \mathcal{C}$$

$$\otimes_\Gamma : |\Gamma,\mathcal{C}| \times |\Gamma^*,\mathcal{C}| \longrightarrow \mathcal{C}.$$

When Γ is a ring (viewed as a category with a single object whose morphisms are right multiplications) and $\mathcal{C}$ is the category of abelian groups, we may identify the categories of additive functors $|\Gamma,\mathcal{C}|$ and $|\Gamma^*,\mathcal{C}|$ respectively with the categories of right and left modules. In this case Hom_Γ and $\otimes_\Gamma$ are just the classical hom and tensor over the ring Γ, which explains the notation.

For simplification we shall assume that Γ is countable and that $\mathcal{C}$ satisfies the following conditions:

a) there is a faithful set-valued functor $S : \mathcal{C} \longrightarrow \mathcal{S}$ which is adjoint to a faithful functor $C : \mathcal{S} \longrightarrow \mathcal{C}$.

b) There exists an internal hom functor

$$\mathrm{Hom}_\mathcal{C} : \mathcal{C}^* \times \mathcal{C} \longrightarrow \mathcal{C}$$

such that $S\,\mathrm{Hom}_\mathcal{C} = \mathrm{hom}_\mathcal{C}$, the set valued hom of $\mathcal{C}$.

c) There exists a tensor product

$$\otimes_\mathcal{C} : \mathcal{C} \times \mathcal{C} \longrightarrow \mathcal{C}$$

which is associative and commutative up to a coherent natural isomorphism.

d) The exponential law holds in $\mathcal{C}$:

$$\mathrm{Hom}_{\mathcal{C}}\,(\mathcal{C}^* \times \mathrm{Hom}_{\mathcal{C}}\,) \cong \mathrm{Hom}_{\mathcal{C}}\,(\otimes^*_{\mathcal{C}} \times \mathcal{C}\,)$$

where $\mathcal{C}$ denotes the identity functor of $\mathcal{C}$.

e) $\mathcal{C}$ is countably complete and cocomplete.

It is easy to see that many algebraic categories satisfy these conditions. We mention in passing that there is a category of topological spaces, namely the category of k-spaces, which has these properties. A space X is a k-space if its topology is determined by all the continuous mappings of compact Hausdorff spaces into X. That is to say, a subset A of X is closed if and only if $f^{-1}A$ is closed for every continuous $f : C \longrightarrow X$ where C is compact Hausdorff. In order for the exponential law to hold it is necessary to use something other than the compact-open topology.

To return to our main discussion, we let $\widetilde{\mathfrak{M}}(\Gamma)$ denote the twisted morphism category of Γ , that is, the category whose objects are the morphisms of Γ and whose morphisms are the commutative squares

$$\begin{array}{ccc} \alpha & \xrightarrow{\ f\ } & \beta \\ {\scriptstyle g}\uparrow & & \downarrow{\scriptstyle h} \\ \alpha' & \xrightarrow[\ f'\ ]{} & \beta' \end{array}$$

In other words (g,h) is a morphism from f to f' if f' = hfg. Now Hom_{Γ} is defined as the composite functor

$$|\Gamma,\mathcal{C}|^* \times |\Gamma,\mathcal{C}| \xrightarrow{\ \tau\ } |\widetilde{\mathfrak{M}}(\Gamma),\mathcal{C}| \xrightarrow{\ \lim\ } \mathcal{C}$$

where $\tau(X,Y)(f) = \mathrm{Hom}_{\mathcal{C}}(X(\mathrm{dom} f),Y(\mathrm{rng} f))$. Similarly $\otimes_{\Gamma}$ is defined by the composition

$$|\Gamma,\mathcal{C}| \times |\Gamma^*,\mathcal{C}| \xrightarrow{\ \sigma\ } |\widetilde{\mathfrak{M}}(\Gamma)^*,\mathcal{C}| \xrightarrow{\ \mathrm{colim}\ } \mathcal{C}$$

where $\sigma(X,Y)(f) = X(\mathrm{dom} f) \otimes_{\mathcal{C}} Y(\mathrm{rng} f)$.

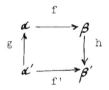

These definitions are not intuitive and it is somewhat more enlightening to describe Hom_{Γ} and $\otimes_{\Gamma}$ in terms of universal properties. For example, when $X \in |\Gamma,\mathcal{C}|$ and $Y \in |\Gamma^*,\mathcal{C}|$ we have for each morphism of Γ, $f : \alpha \longrightarrow \beta$, a commutative diagram in $\mathcal{C}$:

$$\begin{array}{ccccc} & & X(\alpha)\otimes Y(\alpha) & & \\ & \overset{1\otimes Y(f)}{\nearrow} & & \overset{p_\alpha}{\searrow} & \\ X(\alpha)\otimes Y(\beta) & & & & X\otimes_\Gamma Y \\ & \underset{X(f)\otimes 1}{\searrow} & & \underset{p_\beta}{\nearrow} & \\ & & X(\beta)\otimes Y(\beta) & & \end{array}$$

$X \otimes_\Gamma Y$ is universal with respect to this property in the sense that $X \otimes_\Gamma Y$ maps into any other object of $\mathcal{C}$ for which p_α's are defined so that corresponding diagrams exist.

Of course the tensor and hom notation suggest looking for some version of the exponential law in this more abstract context. If we let $\mathrm{Hom}_{\mathcal{C}}(Y^*,Z)$ denote the composite

$$\Gamma \xrightarrow{\quad Y^* \quad} \mathcal{C}^* \xrightarrow{\quad \mathrm{Hom}_{\mathcal{C}}(\ ,Z) \quad} \mathcal{C}$$

for $Y \in |\Gamma^*, \mathcal{C}|$ and $Z \in \mathcal{C}$, then the following holds.

<u>Theorem</u>. For $X \in |\Gamma, \mathcal{C}|$, $Y \in |\Gamma^*, \mathcal{C}|$, and $Z \in \mathcal{C}$,

$$\mathrm{Hom}\ (X \otimes_\Gamma Y, Z) \cong \mathrm{Hom}_\Gamma (X, \mathrm{Hom}_{\mathcal{C}}(Y^*,Z)).$$

(More precisely there is a natural equivalence of functors). There is some tedious work to be done in proving this and therefore I shall not try to give the proof. It is worth observing that $X \otimes_\Gamma Y \cong Y \otimes_{\Gamma^*} X$ and that consequently

$$\mathrm{Hom}_{\mathcal{C}}(X \otimes_\Gamma Y, Z) = \mathrm{Hom}_{\Gamma^*}(Y, \mathrm{Hom}_{\mathcal{C}}(X^*,Z)).$$

To give some examples we shall take Γ to be the category whose objects are the ordered sets

$$\Delta^n = \{0, 1, \ldots, n\}$$

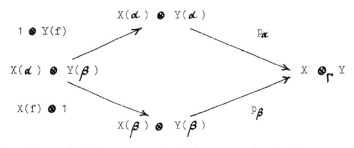

and whose morphisms are the set mappings which preserve the relation $\leq$. We shall call this the <u>category of simplicial prototypes</u> and denote it by Δ. It is well-known that $|\Delta^*, \mathcal{C}|$ for any category $\mathcal{C}$ is just the category of simplicial objects over $\mathcal{C}$. Similarly $|\Delta, \mathcal{C}|$ is just the category of cosimplicial objects over $\mathcal{C}$. For all our examples here we shall take $\mathcal{C}$ to be the category of k-spaces described above. Furthermore we shall fix in $|\Delta, \mathcal{C}|$ an object $|\Delta|$ called the <u>standard simplex</u> for which

$$|\Delta|_n = |\Delta|(\Delta^n) = \{(t_1, t_2, \ldots, t_n) \in I^n \mid 0 \leq t_1 \leq t_2 \leq \ldots \leq t_n \leq 1\}$$

is just the standard n-simplex. Then for any simplicial object over $\mathcal{C}$, say $Y \in |\Delta^*, \mathcal{C}|$, the tensor product $|\Delta| \otimes_\Delta Y = |Y|$ is an object of $\mathcal{C}$ which acts as a geometric realization of Y. Indeed if $Y = \mathcal{S}B$, the total singular complex of B, then $|\Delta| \otimes_\Delta \mathcal{S}B$ is precisely the geometric realization of $\mathcal{S}B$ described by Milnor |2|, and this is our first example.

To obtain a more interesting example we let G denote an associative H-space in the category $\mathcal{C}$ of k-spaces. We shall associate with G a complex $\mathcal{K}G \in |\Delta^*, \mathcal{C}|$ as follows. The n-th component $(\mathcal{K}G)_n$ is just $G^n = G \times \ldots (n) \ldots \times G$ and the faces and degeneracies are given by the formulas:

$$\partial_i(g_1, g_2, \ldots, g_n) = \begin{cases} (g_2, \ldots, g_n) & \text{for } i = 0 \\ (g_1, \ldots, g_i g_{i+1}, \ldots, g_n) & \text{for } i = 1,2,\ldots,n-1 \\ (g_1, \ldots, g_{n-1}) & \text{for } i = n \end{cases}$$

$$s_i(g_1, g_2, \ldots, g_n) = \begin{cases} (e, g_1, \ldots, g_n) & \text{for } i = 0 \\ (g_1, \ldots, g_i, e, g_{i+1}, \ldots, g_n) & \text{for } i=1,2,\ldots,n-1 \\ (g_1, \ldots, g_n, e) & \text{for } i = n \end{cases}$$

where e denotes the unit element of G, which must of course be a strict unit and not just a homotopy unit.

Now the Milgram classifying space |3| for G is just the space $BG = |\Delta| \otimes_\Delta \mathcal{K}G$, which we can also write as $|\mathcal{K}G|$. This is our second example.

Finally we consider an open covering $\mathfrak{U}$ of a k-space X. Suppose $N(\mathfrak{U})$ is the nerve of $\mathfrak{U}$. Then we define a complex $(X, \mathfrak{U})$ with $(X, \mathfrak{U})_n = \bigcup \sup \alpha$ where α runs through the n simplexes of $N(\mathfrak{U})$. Face and degeneracy operators are given in the obvious manner. Now we form the space $|(X, \mathfrak{U})| = |\Delta| \otimes_\Delta (X, \mathfrak{U})$, which we call the <u>space of the covering</u>. It has several interesting properties.

1) If $\mathfrak{U}$ is a refinement of $\mathfrak{V}$, then there is a natural mapping

$$|(X, \mathfrak{U})| \longrightarrow |(X, \mathfrak{V})|$$

2) There is a natural mapping $\beta : |(X,\mathcal{U})| \longrightarrow X$ which is a weak homotopy equivalence.

3) A locally finite covering $\mathcal{U}$ of X is numerable if and only if the mapping β has a cross section.

4) Every mapping of a paracompact Hausdorff space into X factors through β.

5) $|(X,\mathcal{U})|$ has a natural filtration which gives rise to the Mayer-Victoris spectral sequence of the open covering $\mathcal{U}$.

Finally we shall show how to use the space $|(X,\mathcal{U})|$ to obtain the classifying map for a Steenrod fibre bundle over X with $\mathcal{U}$ a coordinate covering. Let E be a bundle with structure group G over X and $\mathcal{U}$ a covering of X by coordinate neighbourhoods. For U_λ, U_μ in $\mathcal{U}$, we have a coordinate function $g_{\lambda\mu} : U_\lambda \cap U_\mu \longrightarrow G$ whenever $U_\lambda \cap U_\mu \neq \emptyset$. These satisfy the 1-cocycle condition: on $U_\lambda \cap U_\mu \cap U_\nu$ we have

$$g_{\lambda\mu} \cdot g_{\mu\nu} = g_{\lambda\nu} .$$

By means of these coordinate functions we can give a morphism of simplicial sets, $(X,\mathcal{U}) \longrightarrow \mathcal{K}G$. This is defined on the n-th component by mappings

$$U_{\lambda_0} \; U_{\lambda_1} \; \ldots \; U_{\lambda_n} \longrightarrow G \times \ldots (n) \ldots \times G$$

given by

$$x \longmapsto (g_{\lambda_0\lambda_1}(x), g_{\lambda_1\lambda_2}(x), \ldots, g_{\lambda_{n-1}\lambda_n}(x))$$

with the convention that $g_{\lambda\lambda}(x) = e$ for all $x \in U_\lambda$.

This mapping $(X,\mathcal{U}) \xrightarrow{\;g\;} \mathcal{K}G$ passes to a mapping of tensor products $|(X,\mathcal{U})| \xrightarrow{\;|g|\;} |\mathcal{K}G| = BG$. If $\mathcal{U}$ is a numerable covering, as remarked in 3) above, there is a mapping $X \xrightarrow{\;\alpha\;} |(X,\mathcal{U})|$ which is a cross section to β. It is easy to verify that the composite

$$X \xrightarrow{\;\alpha\;} |(X,\mathcal{U})| \xrightarrow{\;|g|\;} |\mathcal{K}G| = BG$$

is a classifying mapping for the bundle E over X.

Unfortunately we do not have time to show how other constructions and standard theorems about them can be given in this context, but I hope that these examples, simple as they are, offer some convincing evidence of the usefulness of these ideas in algebraic topology.

REFERENCES

[1] R. J. Milgram, The homology of iterated loop spaces, Annals of Mathematics 2, Vol. 84 (1966), pp. 386-403.

[2] J. Milnor, The geometric realization of a semisimplicial complex, Annals of Mathematics 2, Vol. 65 (1957), pp. 357-362.

[3] N. E. Steenrod, Milgram's classifying space of a topological group, Topology 7 (1968), pp. 349-368.

ON A TYPE OF DIFFERENTIAL HOPF ALGEBRAS

Shôrô Araki

Osaka City University and
Forschungsinstitut für Mathematik ETH, Zürich

The present talk is based on a joint work with Z. Yosimura [3]. The details will be published elsewhere.

1. When we plan to organize a theory of Hopf algebras modelled on mod p K-theory, we will meet with two significant difference from the ordinary theory of Hopf algebras.

The first point is: the classical Hopf algebras modelled on ordinary homology and cohomology of H-spaces is non-negatively graded and can be discussed sometimes by making use of an induction argument on degrees (cf. [7] etc.), but our Hopf algebra is Z_2-graded, so we can not use such arguments. We solve some of these type of problems by using the two filtrations (F-filtration by multiplication and G-filtration by comultiplication) originally due to Browder [4].

The second point is: in the classical theory of Hopf algebras the relation

$$\psi\varphi = (\varphi \otimes \varphi)(1 \otimes T \otimes 1)(\psi \otimes \psi)$$

of Milnor-Moore [7] is very important. But, as I mentioned in my previous paper [1], this relation fails to hold in case of Hopf structures derived from $K^*(\ ;Z_2)$. The existence of Milnor-Moore relation in the classical case is essentially based on the commutativities of external multiplications of ordinary homology and cohomology. But the external multiplication of $K^*(\ ;Z_2)$ is never commutative.

Fortunately the deviation from the commutativity is known [2]. So I want to regard this non-commutativity as a kind of commutativity relation. We put

$$T_1 = (1+d \otimes d)T : K^*(X;Z_2) \otimes K^*(Y;Z_2) \to K^*(Y;Z_2) \otimes K^*(X;Z_2)$$

for two complexes X and Y, where d is the Bockstein homomorphism mod 2. Then we obtain a relation

$$T_1\mu_2 = \mu_2 t^* ,$$

where μ_2 is the external multiplication of $K^*(\ ;Z_2)$ and $t:Y\times X\to X\times Y$ the twisting map. This means that, if we regard T_1 as a kind of twisting morphism, then using T_1 instead of the usual twisting morphism T we have a commutativity of μ_2. Similar situation arises also from the non-commutative multiplications in $K^*(\ ;Z_p)$, p odd prime [2]. Now suppose X to be a finite CW-H-space, then, putting $A = K^*(X;Z_2)$ we have the relation

$$\psi\varphi = (\varphi \otimes \varphi)(1 \otimes T_1 \otimes 1)(\psi \otimes \psi) ,$$

the Milnor-Moore relation in the modified sense.

I will remark one more: since the Bockstein differential appears in T_1, in our situation we must talk of the differential Hopf algebras from the beginning.

2. The above replacement of T by T_1 is quite reasonable. The basic reason to use the twisting morphism T in algebraic topology is as follows: The several tensor products of twisting morphisms and identity maps and their compositions are all the so called signed permutations of tensor factors. And the assignment

$$\mathfrak{S}_n \ni \sigma \to T_\sigma \text{ , signed permutation ,}$$

is a kind of homomorphism of the symmetric group $\mathfrak{S}_n$ with values in certain automorphisms of a category. And we use usually commutation relations in $\mathfrak{S}_n$ as commutation relations of signed permutations. In the theory of Hopf algebras, graded algebras etc., we meet a number of commutation relations of tensor products of twisting morphisms and identity maps; and I believe that all of them come from the commutation relations in $\mathfrak{S}_n$. This is the most basic feature of twisting morphism.

Now we shall observe that the same principle still holds even when we replace T by T_1. To fit also with the case of non-commutative multiplications of $K^*(;Z_p)$, p odd prime, I will slightly generalize T_1.

Let M and N be a Z_2-graded differential module over a field K, and pick an element $\lambda \in K$ and fixed. Put

$$T_\lambda = (1+\lambda \cdot d\sigma \otimes d)T : M \otimes N \to N \otimes M ,$$

where T is the usual twisting morphism, σ is an involution of N such that $\sigma | N_0 = 1$ and $\sigma | N_1 = -1$. We will call T_λ a λ-modified twisting morphism.

Now $M_1, \ldots, M_n$ be Z_2-graded differential modules. Put

$$T_{i,\lambda} = 1 \otimes \ldots \otimes 1 \otimes T_\lambda \otimes 1 \otimes \ldots \otimes 1 , \quad 1 \leq i \leq n-1 ,$$

with T_λ in the i-th tensor factor. $T_{i,\lambda}$ corresponds to the transposition $t_i = (i,i+1) \in \mathfrak{S}_n$. And we can easily check that the following relations hold:

$$\text{i)} \quad (T_{i,\lambda})^2 = 1 \qquad 1 \leq i \leq n-1 ,$$

$$\text{ii)} \quad T_{i,\lambda}T_{i+1,\lambda}T_{i,\lambda} = T_{i+1,\lambda}T_{i,\lambda}T_{i+1,\lambda} \qquad 1 \leq i < n-2 ,$$

$$\text{iii)} \quad T_{i,\lambda}T_{j,\lambda} = T_{j,\lambda}T_{i,\lambda} \qquad i+1 < j .$$

The corresponding relations for t_i's are of course true and make the fundamental relations of $\mathfrak{S}_n$ [5]. So, by expressing each $\sigma \in \mathfrak{S}_n$ as a composition of t_i's, say $\sigma = t_{i_1} \ldots t_{i_s}$ for example, put

$$T_{\sigma,\lambda} = T_{i_1,\lambda} \ldots T_{i_s,\lambda} ,$$

then $T_{\sigma,\lambda}$ is a well-determined map, not depending on the choice of expressions of σ. $T_{\sigma,\lambda}$ will be called by λ-modified permutations of tensor factors. The assignment

$$\mathfrak{S}_n \ni \sigma \to T_{\sigma,\lambda}$$

is now a homomorphism of $\mathfrak{S}_n$, similar to the case of ordinary signed permutations, and we can use commutation relations in $\mathfrak{S}_n$ as those of $T_{\sigma,\lambda}$'s .

3. Now we replace T by T_λ in all formal discussions in the theory of Hopf algebras. For example, let A and B be differential algebras. In the definition of their tensor product we must replace T by T_λ, and obtain $(A \otimes B)_\lambda$, called the λ-modified tensor product, which has a different multiplication from the ordinary tensor product. Similarly is defined the λ-modified tensor product of differential coalgebras. Multiplication φ (or comultiplication ψ) is called to be λ-commutative if it satisfies

$$\varphi T_\lambda = \varphi \quad (\text{or } T_\lambda \psi = \psi) .$$

Finally a differential (quasi) preHopf algebra [1] satisfying the relation

$$\psi\varphi = (\varphi \otimes \varphi)(1 \otimes T_\lambda \otimes 1)(\psi \otimes \psi)$$

is called a λ-modified differential (quasi) Hopf algebra, or simply, (quasi) (d,λ)-Hopf algebra. λ-modified tensor product of two (quasi) (d,λ)-Hopf algebras is again a (quasi) (d,λ)-Hopf algebra.

In the detailed discussions of several properties we use two filtrations of Browder [4] in many ways. Our algebras and coalgebras are defined over a field K, Z_2-graded, augmented and with units. Hence we have a canonical direct sum decomposition

$$A = K \oplus \bar{A} .$$

Since we do not assume the associativity, the definition of two filtrations should be modified. In case of an algebra A, F-filtration is defined by

$$F^oA = A , F^1A = \bar{A} , F^2A = \bar{A}\cdot\bar{A} , F^3A = \bar{A}\cdot(\bar{A}\cdot\bar{A}) + (\bar{A}\cdot\bar{A})\cdot\bar{A} , \ldots$$

In case of coalgebras G-filtration should be defined in a dual way modified.

Since we use the above filtrations to get some results about A itself, we need completeness conditions of filtrations sometimes. If F-filtration (decreasing) is complete, i.e.,

$$\cap F^kA = \{0\} ,$$

the algebra A is called to be semi-connected. Similarly a coalgebra A is called to be semi-connected if

$$\cup G^kA = A .$$

Remark that, when A is (non-negatively) graded and connected, then A is semi-connected.

4. Our first basic property to be established is the so-called Milnor-Moore criterion of coprimitivities of Hopf algebras, [7] Proposition 4.20, in our versions.

Using the two filtrations we obtain the following lemmas.

Lemma 1. Let $f : A \to B$ be a morphism of algebras. f is almost surjective (i.e., f(A) is dense in B by the F-topology) if and only if $Q(f) : Q(A) \to Q(B)$ is surjective.

Lemma 2. Let $f : A \to B$ be a morphism of coalgebras, and assume A to be semi-connected. Then, f is injective if and only if $P(f) : P(A) \to P(B)$ is injective.

Now let A be a quasi (d,λ)-Hopf algebra. Our version of Milnor-Moore relation says that

$$\psi : A \to (A \otimes A)_\lambda$$

is a morphism of differential algebras and

$$\varphi : (A \otimes A)_\lambda \to A$$

is a morphism of differential coalgebras.

We look at the maps

$$\varphi_a = \varphi(\varphi \otimes 1 - 1 \otimes \varphi) : (A \otimes A \otimes A)_\lambda \to A$$

and

$$\varphi_c = \varphi(T_\lambda - 1) : (A \otimes A)_\lambda \to A ,$$

then we see easily that $\operatorname{Im} \varphi_a$ and $\operatorname{Im} \varphi_c$ are <u>coideals</u> of A (with respect to the coalgebra structure of A). So $\operatorname{coker} \varphi_a$ and $\operatorname{coker} \varphi_c$ are quotient coalgebras and the projection $\pi : A \to \operatorname{coker} \varphi_a$ is a morphism of coalgebras. Since $\operatorname{Im} \varphi_a \subset F^2 A$ we have an induced map $f : \overline{\operatorname{coker} \varphi_a} \to Q(A)$ making the following diagram commutative:

$$\begin{array}{ccccc} P(A) & \xrightarrow{\ i\ } & \bar{A} & \xrightarrow{\ j\ } & Q(A) . \\ \Big\downarrow{\scriptstyle P(\pi)} & & \Big\downarrow{\scriptstyle \bar{\pi}} & \nearrow{\scriptstyle f} & \\ P(\operatorname{coker} \varphi_a) & \longrightarrow & \overline{\operatorname{coker} \varphi_a} . & & \end{array}$$

Assume that A is coprimitive and semi-connected as coalgebra. Then $\nu = ji$ is injective; thus by the above diagram we see that $P(\pi)$ is injective. Then, by Lemma 2 we see that π is injective, i.e., $\varphi_a = 0$; similarly we see that $\varphi_c = 0$. Thus we obtain

<u>Proposition 3</u>. Let A be a quasi (d,λ)-Hopf algebra which is semi-connected as coalgebra. If A is coprimitive then the multiplication φ is associative and λ-commutative.

By a discussion dual to the above we obtain

<u>Proposition 4</u>. Let A be a quasi (d,λ)-Hopf algebra which is semi-connected as algebra. If A is primitive then the comultiplication ψ is associative and λ-commutative.

5. The next thing we must establish is another condition of coprimitivity in case of char $K = p \neq 0$. Milnor-Moore [7] considered the map

$$x \to x^p \ , \ x \in \bar{A} \ ,$$

and their condition is that this map should be zero. This formulation of the condition seems to be inconvenient by two reasons: i) the above map is generally not K-linear (even though additive), and to treat the matter of Hopf algebras within the category of linear spaces and linear maps over a field K , this formulation seems to be not the best; ii) our λ-commutativity is not the usual commutativity, so we do not know whether the above map makes sense in advance.

We make the following construction. Let A be a proper differential algebra (or coalgebra) over a field K and $\lambda \in K$. Suppose $p = \text{char } K \neq 0$. λ-modified cyclic permutation

$$C_\lambda : (A^{\otimes p})_\lambda \to (A^{\otimes p})_\lambda$$

is a morphism of differential algebra (or coalgebra) such that $C_\lambda^p = 1$. Put

$$\Delta_\lambda = 1-C_\lambda \quad \text{and} \quad \Sigma_\lambda = 1+C_\lambda+\ldots+C_\lambda^{p-1} \ ;$$

and define

$$\Phi_\lambda A = \ker \Sigma_\lambda / \text{Im } \Delta_\lambda \quad \text{and} \quad \Psi_\lambda A = \ker \Delta_\lambda / \text{Im } \Sigma_\lambda \ .$$

We can see that $\Psi_\lambda A$ (or $\Phi_\lambda A$) has an induced structure of differential algebra (or coalgebra). We can also prove that the differentials of $\Phi_\lambda A$ and $\Psi_\lambda A$ vanish.

Since $\Sigma_\lambda = \Delta_\lambda^{p-1}$ we have inclusions

$$\ker \Delta_\lambda \subset \ker \Sigma_\lambda \quad \text{and} \quad \text{Im } \Sigma_\lambda \subset \text{Im } \Delta_\lambda \ ,$$

which induce a natural map

$$\Psi_\lambda A \to \Phi_\lambda A \ .$$

In fact it turns out to be an isomorphism. So, when A is a (quasi) (d,λ)-Hopf algebra, identifying by the above natural isomorphism, $\Psi_\lambda A = \Phi_\lambda A$ becomes a (quasi) Hopf algebra, which we call a <u>derived Hopf algebra</u> of A .

Now assume φ (or ψ) is associative and λ-commutative. (p-1)-fold multiplication (or comultiplication) induces a morphism of Hopf algebras

$$\xi_\lambda : \Phi_\lambda A \to A \quad (\text{or} \quad \eta_\lambda : A \to \Psi_\lambda A \ ) \ .$$

We put $\bar{\xi}_\lambda = \xi_\lambda | \overline{\Phi_\lambda A}$ and $\bar{\eta}_\lambda = \eta_\lambda | \bar{A}$.

We formulate the conditions for coprimitivity and primitivity of A (in case of $\operatorname{char} K = p \neq 0$) as follows:

<cp> the multiplication φ is associative and λ-commutative, and $\bar{\xi}_\lambda$ is a zero map,

<p> the comultiplication ψ is associative and λ-commutative, and $\bar{\eta}_\lambda$ is a zero map.

We can prove

<u>Proposition 5</u>. The properties <cp> and <p> of A are hereditary to $H(A)$.

6. Now we have

<u>Theorem 1</u>. Let $\lambda \in K$, $\operatorname{char} K = p \neq 0$, and A be a quasi (d,λ)-Hopf algebra over K .

i) Assume that A is semi-connected as coalgebra. If A is coprimitive then A satisfies <cp> .

ii) Assume that A and $\Psi_\lambda A$ is semi-connected as algebra. If A is primitive then A satisfies <p> .

As to the proof we remark only that the proof of $\bar{\xi}_\lambda = 0$ (or

$\bar{\eta}_\lambda = 0$) is quite parallel to the proof of Proposition 3 (or 4).

As a converse to the above we have

Theorem 2. Let $\lambda \in K$, char $K = p \neq 0$, and A be a quasi (d,λ)-Hopf algebra over K. Suppose that p is odd or that $p = 2$ and $\lambda d = 0$.

i) Assume that A is semi-connected as algebra. If A satisfies <cp> then A is coprimitive.

ii) Assume that A is semi-connected as coalgebra. If A satisfies <p> then A is primitive.

The proof of Theorem 2 is much more difficult and massy. In Theorem 2 an important case of $p = 2$ and $\lambda d \neq 0$ is excluded. But I doubt any proof could exist in this case. Maybe counter examples exist even though I have none at hand.

7. Let A be a (quasi)(d,λ)-Hopf algebra over a field K, char $K = p \neq 0$. F- and G-filtration of A define spectral sequences

$$\{E_r(A)\ ,\ r \geq 0\} \quad \text{and} \quad \{{}_rE(A)\ ,\ r \geq 0\}$$

in which all terms are graded and connected; $E_0(A)$ and ${}_0E(A)$ are (quasi)(d,λ)-Hopf algebras for the same $\lambda \in K$; $E_r(A)$ and ${}_rE(A)$ are (quasi) differential Hopf algebras (i.e., $\lambda = 0$). As usual $E_0(A)$ is primitive and ${}_0E(A)$ is coprimitive (cf. [1] and [4]). Now by Proposition 5, Theorems 1 and 2, we obtain (without exclusions)

Proposition 6. $E_r(A)$ are primitive and ${}_rE(A)$ are coprimitive for all $r \geq 0$.

One might now expect the biprimitivity of ${}_rE(E_s(A))$ or $E_r({}_sE(A))$ for all (s,t). This is certainly true when p is odd or $p = 2$ and $\lambda d = 0$. In case $p = 2$ and $\lambda d \neq 0$ we have a trouble in the first term ${}_0E(E_0A)$ or $E_0({}_0E(A))$. (The second half of Theorem 6 of [3] is incorrect). Nevertheless, assume that A is finite dimensional with associative and λ-commutative multiplication, then we can prove Borel structure theorem for E_oA such that the generators are d-stable. Using this we can prove that ${}_oE(E_oA)$ is primitive, hence biprimitive; therefore ${}_rE(E_sA)$ is biprimitive for all (r,s). Thus, in a parallel way to [4] we obtain

Theorem 3. Let X be a finite CW-H-complex. For any prime p there exists a spectral sequence of which every term is a biprimitive (d,λ)-Hopf algebra, the first term is a biprimitive form of $K^*(X;Z_p)$, i.e., ${}_oE(E_oK^*(X;Z_p))$, and the last term is a biprimitive form of $(K^*(X)/\mathrm{Tors}) \otimes Z_p$.

Using the above Theorem we can give a short and unified proof of Hodgkin's Theorem on $K^*(G)$ with simply connected G. Furthermore, using the above Theorem and Karoubi's description of mod p K-theory [6], we can determine $K^*(SO(n))$.

Theorem 4. i) $K^*(SO(2n+1)) = \Lambda_Z(\beta(\lambda_1),\ldots,\beta(\lambda_{n-1}),a_n)$
$\oplus\, Z_{2^n}\{\eta_{2n+1}\} \otimes \Lambda_Z(\beta(\lambda_1),\ldots,\beta(\lambda_{n-1}))$

with relations: $\eta\cdot a_n = 0$ and $\eta^2 = 2\eta$;

ii) $K^*(SO(2n)) = \Lambda_Z(\beta(\lambda_1),\ldots,\beta(\lambda_{n-2}),a_n^+,b_n)$
$\oplus\, Z_{2^{n-1}}\{\eta_{2n}\} \otimes \Lambda_Z(\beta(\lambda_1),\ldots,\beta(\lambda_{n-2}),b_n)$

with relations: $\eta\cdot a_n^+ = 0$ and $\eta^2 = 2\eta$,

where a_n, a_n^+ and b_n are elements of degree 1 represented by maps
$a_n : x \to \Delta(\bar{x})^2 \in U(2^n)$, $a_n^+ : x \to \Delta^+(\bar{x})^2 \in U(2^{n-1})$ and
$b_n : x \to \Delta^+(\bar{x})\,\bar{\Delta}(\bar{x}) \in U(2^{n-1})$, $\bar{x} \in \pi^{-1}(x)$, $\pi : \mathrm{Spin}(m) \to SO(m)$,
and η is the reduced line bundle defined by the covering π.

The same results are also obtained by Hodgkin, Lam, Held and Suter by different methods.

REFERENCES

[1] S. Araki, Hopf structures attached to K-theory: Hodgkin's theorem Ann. of Math., 85 (1967), 508-525.

[2] S. Araki and H. Toda, Multiplicative structures in mod q cohomology theories, I and II. Osaka J. Math., 2 (1965), 71-115 and 3 (1966), 81-120.

[3] S. Araki and Z. Yosimura, On a certain type of differential Hopf algebras. Proc. Japan Acad., 46 (1970), 332-336.

[4] W. Browder, On differential Hopf algebras. Trans. Amer. Math. Soc., 107 (1963), 153-176.

[5] L.E. Dickson, Linear groups with an exposition of the Galois field theory. Leipzig, Teubner, 1901.

[6] M. Karoubi, Algèbres de Clifford et K-théorie. Ann. Ec. Norm. Sup. (4) 1 (1968), 1-90.

[7] J.W. Milnor and J.C. Moore, On the structure of Hopf algebras. Ann. of Math., 81 (1965), 211-264.

POLYNOMIAL ALGEBRAS OVER THE ALGEBRA OF COHOMOLOGY OPERATIONS

Norman Steenrod
Churchill College, Cambridge, G.B.

Introduction. The problem of constructing finitely generated polynomial algebras over the algebra $\mathcal{A}(p)$ of reduced power operations is important in the study of classifying spaces because their cohomologies frequently have this form. The problem is difficult because the dimensions of the generators cannot be assigned at random. For example Adem [1] observed that, if p is an odd prime, and there is just one generator of dimension 2n, then n divides $(p-1)p^h$ for some h. Thomas [4] restricted himself to the case p = 2, and showed that the existence of a generator in dimension n requires generators in every dimension k such that $\binom{k-1}{n-k} \equiv 1 \bmod 2$. For example, if n = 17, there must be generators in all dimensions from 2 to 16. The strength of this result has been amply demonstrated by Sugawara and Toda [3].
Various unsuccessful efforts to generalize Thomas's results to $p > 2$ leads one to search for examples of polynomial algebras over $\mathcal{A}(p)$ to provide tests of possible formulations of the generalization. Just about the only known ones are a few cohomology algebras of classifying spaces that have been computed. The direct but naive method of specifying generators in certain dimensions, and then the action of the generators $\{P^i\}$ of $\mathcal{A}(p)$ on them, leads to the problem of verifying in a finite number of steps that the resulting polynomial algebra is indeed an algebra over $\mathcal{A}(p)$. In this paper we supply a method for doing this, and we apply it to a non-trivial example having two generators in degrees 4 and $2(p+1)$. A main feature of this work is the derivation of a set of relations among the Adem relations.

2. The free associative Hopf algebra $\bar{\mathcal{A}}(p)$. For a prime p, let M be the graded Z_p-module spanned by the elements $\{P^i\}$, i = 1,2,..., where $\deg(P^i) = 2i(p-1)$. (When p = 2, replace P^i by Sq^i of degree i.) Let $\bar{\mathcal{A}}(p) = \otimes M$, so that $\bar{\mathcal{A}}$ is the free associative algebra generated by $\{P^i\}$. The coproduct $\bar{\psi} : \bar{\mathcal{A}} \to \bar{\mathcal{A}} \otimes \bar{\mathcal{A}}$ is defined by the Cartan formula $\bar{\psi} P^k = \sum_{i=0}^{k} P^i \otimes P^{k-i}$ (by convention, P^0 is the unit), and then $\bar{\mathcal{A}}$ is a Hopf algebra. Identifying P^i with the reduced p^{th} power operation $H^n(X) \to H^{n+2i(p-1)}(X)$ induces a morphism of Hopf algebras $\lambda : \bar{\mathcal{A}} \to \mathcal{A}$. For integers a,b such that $0 < a < pb$, define $R_{a,b}$ by

$$(2.1)\quad R_{a,b} = P^a \otimes P^b - \sum_{t=0}^{[a/p]} (-1)^{a+t} \binom{(p-1)(b-t)-1}{a-pt} P^{a+b-t} \otimes P^t,$$

then Ker λ is the ideal generated by the Adem relations $\{R_{a,b}\}$.

A basis for the first three terms of $\bar{\mathcal{A}}$, namely, $Z_p + M + M \otimes M$ is given by 1, all P^i, and all $P^i \otimes P^j$. We call $P^a \otimes P^b$ <u>admissible</u> if $a \geq pb$. Replacing each inadmissible $P^a \otimes P^b$ by $R_{a,b}$ yields a new basis for $Z_p + M + M \otimes M$ called the <u>admissible</u> basis. Let V be the subspace spanned by the $R_{a,b}$, and U the subspace spanned by 1, all P^i, and admissible $P^a \otimes P^b$. Since admissible monomials are independent in $\mathcal{A}$, it follows that λ is monic on U, and $\lambda V = 0$.

2.2. <u>Lemma</u>. The coproduct $\bar{\psi}\, R_{a,b}$ in $\bar{\mathcal{A}} \otimes \bar{\mathcal{A}}$ can be written

$$\bar{\psi}\, R_{a,b} = \sum_{i,j} (\alpha_{i,j} \otimes R_{i,j} + R_{i,j} \otimes \beta_{i,j})$$

for some choice of the α's and β's in $\bar{\mathcal{A}}$, and the sum is taken only over inadmissible i,j such that $i \leqq a$, $j \leqq b$.

<u>Example</u>. For $p = 2$, $\bar{\psi}\, R_{3,2} = \bar{\psi}\,(Sq^3\, Sq^2) = Sq^3\, Sq^2 \otimes 1 + (Sq^2\, Sq^2 + Sq^3\, Sq^1) \otimes Sq^1 + (Sq^1\, Sq^2 + Sq^3) \otimes Sq^2 + Sq^1\, Sq^1 \otimes Sq^2 Sq^1$

+ terms symmetric to these $= R_{3,2} \otimes 1 + R_{2,2} \otimes Sq^1 + R_{1,2} \otimes Sq^2 + R_{1,1} \otimes Sq^2\, Sq^1$ + the symmetric terms.

<u>Proof</u>. By definition, $\bar{\psi}$ maps each $P^a \otimes P^b$ into

$$(U + V) \otimes (U + V) = U \otimes U + U \otimes V + V \otimes U + V \otimes V,$$

hence the same is true of $R_{a,b}$. Since admissible monomials are independent in $\mathcal{A}$, λ is monic on U, and $\lambda V = 0$; hence $\lambda \otimes \lambda$ is monic on $U \otimes U$, and zero on $U \otimes V + V \otimes U + V \otimes V$. Since $(\lambda \otimes \lambda)\bar{\psi}\, R_{a,b} = \psi\, \lambda\, R_{a,b} = 0$, it follows that $\bar{\psi}\, R_{a,b}$ lies in $U \otimes V + V \otimes U + V \otimes V$, hence $\bar{\psi}\, R_{a,b}$ is expressible as a sum of the form given in 2.2, but it remains to show that the restrictions $i \leq a$, $j \leq b$ are fulfilled. The procedure for computing $\bar{\psi}\, R_{a,b}$ in terms of admissible bases for U, V is to take $\bar{\psi}$ of each term of 2.1, apply the Cartan formula to each factor, multiply out, and, whenever an inadmissible $P^i P^j$ occurs, substitute using 2.1. After all cancellations the desired form is obtained. Thus an $R_{i,j}$ appears only when an inadmissible $P^i P^j$ occurs in the $\bar{\psi}$ of a term of 2.1. But $\bar{\psi}\,(P^a \otimes P^b) = \sum_{i,j=(1,1)}^{(a,b)} P^i P^j \otimes P^{a-i} P^{b-j}$, so these satisfy $i \leq a$, $j \leq b$, and $a - i \leq a$, $b - j \leq b$. Now $\bar{\psi}(P^{a+b-t} \otimes P^t)$ has terms of the form $P^i P^j \otimes P^{a+b-t-i} P^{t-j}$ where

$$0 \leq i \leq a + b - t, \quad 0 \leq j \leq t, \text{ and } pt \leq a < pb.$$

These inequalities imply that both $j \leq b$ and $t - j \leq b$. If $P^i P^j$ is inadmissible, we have $i < pj \leq pt \leq a$ as required. If

$P^{a+b-t-i}P^{t-j}$ is inadmissible we have

$$a + b - t - i < p(t-j) = pt - pj \leq a - pj \leq a. \quad \square$$

3. Unstable modules and algebras over $\bar{\mathcal{A}}$. A graded module N over $\bar{\mathcal{A}}$ is said to be unstable if $P^i x = 0$ whenever $2i > \deg x$ ($i > \deg x$ for $p=2$). It is readily verified that submodules and quotient modules of unstable modules are unstable.

Recall that, for modules N, N' over a Hopf algebra $\bar{\mathcal{A}}$, the tensor product $N \otimes N'$, that is naturally an $\bar{\mathcal{A}} \otimes \bar{\mathcal{A}}$ - module, becomes an $\bar{\mathcal{A}}$-module by means of the coproduct $\bar{\mathcal{A}} \to \bar{\mathcal{A}} \otimes \bar{\mathcal{A}}$.
Thus for $x \in N_q$, $y \in N'_r$, we have

(3.1) $P^k(x \otimes y) = \sum_{i=0}^{k} P^i x \otimes P^{k-i} y.$

If $2k > q + r$, either $2i > q$ or $2k - 2i > r$. Hence a tensor product of unstable $\bar{\mathcal{A}}$-modules is unstable.

Now let A be a graded algebra over the Hopf algebra $\bar{\mathcal{A}}$, i.e. the multiplication $A \otimes A \to A$ is a map of $\bar{\mathcal{A}}$-modules (see [2]). We say that A is unstable as an algebra if it is unstable as a module, and $P^k x = x^p$ when $2k = \deg x$ ($Sq^k x = x^2$ when $k = \deg x$).

3.2. Lemma. Let A be a commutative algebra over $\bar{\mathcal{A}}$ such that the instability conditions hold for each element of a set S of generators of A, then A is an unstable algebra over $\bar{\mathcal{A}}$.

Proof. Let S^n denote the set of n-fold products of elements of S. Assume inductively that the instability conditions hold for all elements of S^{n-1}. Let $z = xy$ where $x \in S$, $y \in S^{n-1}$ have degrees q, r respectively. Now $P^k(xy) = \sum_{i=0}^{k} (P^i x)(P^{k-i} y)$, so $2k > q + r$ implies either $2i > q$ or $2k - 2i > r$, then $P^k(xy) = 0$. If $2k = q + r$ and both q, r are odd, then again $2i > q$ or $2k - 2i > r$, so again all terms are zero, in agreement with $x^p = 0$ and $y^p = 0$ because A is commutative and q, r are odd. If $q = 2\ell$, $r = 2m$ are even, and $2k = q + r$, then, for all but one term, we have $2i > q$ or $2k - 2i > r$ and the term is zero; the non-zero term has $i = \ell$ and $k-i = m$, then $P^k(xy) = (P^\ell x)(P^m y) = x^p y^p = (xy)^p$. This completes the proof that the instability conditions hold for all products of generators. Since raising to a pth power is a linear operation, the instability conditions hold also for linear combinations of these products. $\square$

3.3. Lemma. Let A be a commutative unstable algebra over $\bar{\mathcal{A}}$, and let $x \in A_{2k}$. Then $P^s P^k x = P^s x^p = 0$ for every s not a multiple of p, and $P^{pt} P^k x = (P^t x)^p$.

Proof. By the Cartan formula $P^s x^p = \sum (P^{i_1}x)\ldots(P^{i_p}x)$ where the sum is taken over all sequences $i_1, \ldots, i_p$ of non-negative integers whose sum is s. If the terms of such a sequence are not all equal, their cyclic permutations determine p distinct terms of the sum having the same value, hence adding to zero. It follows that $P^s x^p$ is zero unless s has the form pt, and then the sum reduces mod p to the one term $(P^t x)\ldots(P^t x)$ where all the indices are equal. $\square$

3.4. Lemma. If A and x are as in 3.3 and $b \geq k$, then $R_{a,b}\, x = 0$.

Proof. Suppose first that $b > k$. Instability implies that $P^b x = 0$, hence $(P^a \otimes P^b)\, x = 0$. If $0 \leq t \leq [a/p]$, then

$$(3.5)\quad \deg P^t x = 2k + 2t(p-1) < 2b + 2a - 2t = 2(a+b-t);$$

hence instability implies that $(P^{a+b-t} \otimes P^t)x = 0$. Thus each term on the right of 2.1 is zero on x. Suppose now that $b = k$, and that a is not of the form pr. By 3.3, $(P^a \otimes P^k)\, x = 0$. Also the calculation 3.5 is still valid because $tp < a$. So again all terms of 2.1 are zero on x. Finally let $b = k$ and $a = pr$. By 3.3, $(P^{pr} \otimes P^k)x = (P^r x)^p$. The calculation 3.5 remains valid for $0 \leq t < a/p = r$, so all terms of the sum are zero on x except possibly when $t = r$, and then we obtain

$$(-1)^{pr+r} \binom{(p-1(b-r)-1}{0} P^{k+r(p-1)} \otimes P^r x = (P^r x)^p. \quad \square$$

4. Verifying Adem relations. Starting with an unstable algebra A over $\bar{\mathcal{a}}$ that is zero in odd degrees when $p > 2$ (e.g. a polynomial algebra), to show that the $\bar{\mathcal{a}}$-action induces an $\mathcal{a}$-action it suffices to show that the Adem ideal Ker λ acts trivially on A. Since the $R_{a,b}$ generate the ideal it suffices to show that each $R_{a,b}$ acts trivially on A.

4.1. Lemma. Let S be a set of generators of A, and $0 < a < pb$, If $R_{i.j}$ acts trivially on S for all $0 < i \leq a$, $0 < j \leq b$, $i < pj$, then $R_{a,b}$ acts trivially on A.

Proof. Let S^n be the set of n-fold products of elements of S. Assume, inductively, that the $R_{i.j}$ as above are trivial on S^{n-1}. Let $x \in S$ and $y \in S^{n-1}$, and let k, ℓ be such that $0 < k \leq a$, $0 < \ell \leq b$, and $k < p\ell$. Then $R_{k,\ell}$ is defined, and $\bar{\psi}\, R_{k,\ell}$ has the form 2.2. Therefore

$$R_{k,\ell}(xy) = \sum ((\alpha_{i,j}x)\,(R_{i,j}y) + (R_{i,j}x)\,(\beta_{i,j}y)).$$

Each term is zero by the inductive hypothesis, and this implies the inductive hypothesis for S^n. It follows now that $R_{a,b}$ acts trivially on every product of generators, and then on their linear combinations.

4.2. <u>Summary</u>. Our original problem may be formulated precisely as follows. Let $A = Z_p[x_1, \ldots x_m]$ be the polynomial algebra on $x_1, \ldots, x_m$ of positive degrees (all even if $p > 2$). For $k = 1, \ldots, m$ and $i = 1, 2, \ldots,$ let $f_{k,i}$ be a definite polynomial in A of degree $2i(p-1) + \deg x_k$ when $p > 2$ ($i + \deg x_k$ when $p = 2$). The problem is: when do the formulas $P^i x_k = f_{k,i}$ define an $\mathcal{A}$-action in A such that A is an unstable algebra over $\mathcal{A}$ as a Hopf algebra?

The first point is that, without further restrictions, they define an $\bar{\mathcal{A}}$-action in A so that A is an algebra over $\bar{\mathcal{A}}$ as a Hopf algebra. To see this, note that each monomial in the generators has a unique form up to permutations of the factors; since the Cartan formula is commutative and associative, it follows that the operations P^i extend uniquely from generators to monomials. Since monomials form a Z_p-basis for A, each P^i extends uniquely over A.

If the polynomials $f_{k,i}$ are chosen so that $f_{k,i} = 0$ for $2i > \deg x_k$ and $f_{k,i} = x_k^p$ for $2i = \deg x_k$, it follows from 3.2, that A is an unstable algebra over $\bar{\mathcal{A}}$. There remains the problem of showing that each $R_{a,b}$ is zero on A. According to 4.1 this follows from showing that $R_{i,j}$ is zero on each generator for $i \leqq a$, $j \leqq b$. Thus we have only to show that every $R_{i,j}x_k = 0$. According to 3.4, the instability properties imply these relations for all i,j,k such that $i < pj$ and $2j \geqq \deg x_k$. Thus we are left with the finite number of cases to check where $i < pj$ and $2j < \deg x_k$. In the next few sections we show how to reduce the number of cases substantially to what we believe is a minimal set of cases.

5. <u>Relations among relations</u>. We proceed to derive certain linear identities in the algebra $\bar{\mathcal{A}}$ involving the $R_{a,b}$ and the products $P^i \otimes R_{j,k}$ and $R_{i,j} \otimes P^k$. These are derived essentially from simpler identities such as $2P^2 = R_{1,1} + P^1 \otimes P^1$. An example will clarify the general case.

Let $p > 3$. Now $P^1 \otimes P^2 \otimes P^1$ is <u>doubly</u> inadmissible in that both the first two and the second two factors are inadmissible. We can reduce it to admissible form in two ways. Beginning with the first two factors, we have

$$P^1 \otimes P^2 \otimes P^1 = R_{1,2} \otimes P^1 + 3P^3 \otimes P^1 = R_{1,2} \otimes P^1 + 3R_{3,1} + 12P^4.$$

Starting with the second two factors gives

$$P^1 \otimes P^2 \otimes P^1 = P^1 \otimes R_{2,1} + 3P^1 \otimes P^3 = P^1 \otimes R_{2,1} + 3R_{1,3} + 12P^4.$$

Notice that the same admissible form $12\,P^4$ is obtained in both cases. Subtracting the two relations gives

$$3\,R_{3,1} = 3\,R_{1,3} + P^1 \otimes R_{2,1} - R_{1,2} \otimes P^1$$

We contract this to a <u>congruence</u> $R_{3,1} \equiv R_{1,3}$ modulo the subspace spanned by the products $P^i \otimes R_{j,k}$ and $R_{j,k} \otimes P^i$ $(i > 0)$. Thus, if we know that $R_{1,2}$, $R_{2,1}$ and $R_{1,3}$ act trivially on an $\bar{\mathcal{A}}$-algebra, it follows that $R_{3,1}$ does also.

5.1. <u>Theorem</u>. Let a,b,c be positive integers such that $a < pb$, and $a + b < pc$. Then we have the following congruence in $\bar{\mathcal{A}}$

$$\binom{(p-1)b-1}{a} R_{a+b,c} \equiv$$

$$(-1)^{a+b}\binom{(p-1)c-1}{b} R_{a,b+c} - \sum_{s=1}^{[a/p]} \binom{(p-1)(b-s)-1}{a-ps}\binom{(p-1)c-1}{s} R_{a+b-s,c+s}.$$

<u>Proof</u>. We shall abbreviate $P^i \otimes P^j \otimes P^k$ by $[i,j,k]$ and $R_{i,j} \otimes P^k$ by $[R_{i,j},k]$, etc. Since $[a,b,c]$ is doubly inadmissible we can reduce it in two ways to the same admissible form using Adem relations; subtracting these two relations gives the above congruence. Starting with the first two factors and the relation $R_{a,b}$, we obtain

$$(1)\quad [a,b,c] = [R_{a,b},c] + (-1)^a\binom{(p-1)b-1}{a}[a+b,c]$$
$$+ \sum_{r=1}^{a/p} (-1)^{a+r}\binom{(p-1)(b-r)-1}{a-pr}[a+b-r,r,c]$$

Since $a+b < pc$, $[a+b,c]$ is inadmissible. Since $r \leqq a/p < b < pc$, $[r,c]$ is inadmissible for $r = 1,\dots,[a/p]$. Therefore reductions are in order for $[a+b,c]$ and all of the $[a+b-r,r,c]$; these are as follows (binomials whose precise forms are of no interest are indicated by greek letters).

$$(2)\quad [a+b,c] = R_{a+b,c} + \alpha\, P^{a+b+c} + \sum_{s=1}^{(a+b)/p} \alpha_s [a+b+c-s,s].$$

$$(3)\quad [a+b-r,r,c] = [a+b-r,R_{r,c}] + (-1)^r\binom{(p-1)c-1}{r}[a+b-r,r+c]$$
$$+ \sum_{s=1}^{r/p} \beta_s\,[a+b-r,r+c-s,s]\,.$$

On the right of (2) all terms are automatically admissible, no further reduction is possible. On the right of (3) all but the first are inadmissible. To see this note that $ps \leqq r$ implies $ps < (p+1)r$; adding this to $a + b < pc$ and rearranging gives $a + b - r < p(r+c-s)$ as required. Reductions of these terms are as follows.

$$(4)\quad [a+b-r,r+c] = R_{a+b-r,r+c} + \gamma\, p^{a+b+c} + \sum_{t=1}^{(a+b-r)p} \gamma_t\, [a+b+c-t,t].$$

$$(5)\quad [a+b-r,r+c-s,s] = [R_{a+b-r,r+c-s},s] + \delta\, [a+b+c-s,s] + \sum_{t=1}^{(a+b-r)/p} \delta_t\, [a+b+c-s-t,t,s].$$

All terms on the right of (4) are automatically admissible. On the right of (5), the term $[a+b+c-s,s]$ is admissible because $(p+1)s \leq (p+1)\, r/p \leq (p+1)\, a/p^2 \leq a$. The terms $[a+b+c-s-t,t,s]$ are inadmissible for $1 \leq t < ps$; a reduction of such a term is

$$(6)\quad [a+b+c-s-t,t,s] = [a+b+c-s-t,R_{t,s}] + \varepsilon\, [a+b+c-s-t,s+t] + \sum_{u=1}^{t/p} \varepsilon_t\, [a+b+c-s-t,s+t-u,u].$$

Since $pt \leq a + b - r$ and $ps \leq r$, we have $p(s+t) \leq a + b < pc$, hence $c - s - t > 0$, and therefore $p(s+t) \leq a + b + c - s - t$. It follows that all terms of (6) are admissible, and no further reductions are possible. Thus the first reduction of $[a,b,c]$ is obtained by substituting (6) in (5), then (4) and (5) in (3), and then (2) and (3) in (1).

The second reduction is obtained by applying a relation to the inadmissible pair b,c.

$$(7)\quad [a,b,c] = [a,R_{b,c}] + (-1)^b \binom{(p-1)c-1}{b} [a,b+c] + \sum_{r=1}^{b/p} (-1)^{b+r} \binom{(p-1)(c-r)-1}{b-pr} [a,b+c-r,r].$$

Since $a < pb$, $[a,b+c]$ is inadmissible. Since $r \leq b/p < c$ and $a < pb$, we have $a < p\,(b+c-r)$, so $[a,b+c-r,r]$ is inadmissible. Reductions of these terms are

$$(8)\quad [a,b+c] = R_{a,b+c} + \lambda\, p^{a+b+c} + \sum_{s=1}^{a/p} \lambda_s\, [a+b+c-s,s].$$

$$(9)\quad [a,b+c-r,r] = [R_{a,b+c-r},r] + \mu\, [a+b+c-r,r] + \sum_{t=1}^{a/p} \mu_t\, [a+b+c-r-t,t,r].$$

All terms on the right side of (8) are automatically admissible. Since $r \leq b/p < c$, we have $pr + r < b + c$, therefore $[a+b+c-r,r]$ is always admissible. But $[a+b+c-r-t,t,r]$ is not for $1 \leq t < pr$; its reduction gives

$$(10)\quad [a+b+c-r-t,t,r] = [a+b+c-r-t,R_{t,r}] + \nu [a+b+c-r-t,r+t] + \sum_{u=1}^{t/p} \nu_t [a+b+c-r-t,r+t-u,u]$$

Since $p(r+t-u) \leqq pr + pt \leqq b + a < pc$, we have $c - r - t > 0$, hence $p(r+t-u) \leqq a + b + c - r - t$. Therefore all terms of (10) are admissible. Thus the second reduction of $[a,b,c]$ to admissible form is obtained by substituting (10) in (9), and then (8) and (9) in (7).

Since the reduction of $[a,b,c]$ to admissible form is unique in the algebra $\mathcal{A}$, it follows that the two reductions coincide on all terms not involving an $R_{i,j}$. Subtracting the two reductions gives us a relation among the R's. Deleting terms of the form $R_{ij} \otimes P^k$ and $P^k \otimes R_{ij}$ yields the stated linear congruence in the R's. □

6. Reduction to a minimal set of relations. The formula of 5.1 we shall denote by $F(a,b,c)$. We shall write frequently $R(i,j)$ for the relation $R_{i,j}$.

6.1. Definition. A relation $R(h,k)$ is called reducible if $R(h,k) \equiv \sum_{i=1}^{h-1} a_i R(h-i,k+i)$ for some integers a_i.

6.2. Theorem. The following reductions hold:

(i) $R(1,mp-1) \equiv 0$ for $m = 2,3,\ldots.$

(ii) If a is not a power of p, then $R(a,b)$ is reducible.

(iii) $R(ap,bp)$ is reducible whenever $R(a,b)$ is reducible.

(iv) $R(p,b)$ is reducible if b is not 2, nor $1 + p^s-p$ for $s = 2,3,\ldots,$ nor $(t + kp)p$ for $t = 1,2,\ldots,p-2$ and $k = 0,1,2,\ldots.$

(v) $R(p^s,k)$ is reducible if k is not divisible by p^{s-1}, or if $k = bp^{s-1}$ and b is restricted as in (iv).

The rest of this section is devoted to the proofs of these statements. When a is restricted to $0 < a < p$, then $[a/p] = 0$, hence $F(a,b,c)$ becomes

$$(1)\quad \binom{p-b_0-1}{a} R(a+b,c) \equiv (-1)^{a+b} \binom{(p-1)c-1}{b} R(a,b+c).$$

The p-adic expansion of b is denoted by $\sum_i b_i p^i$, and the mod p values of the binomials are computed using the rule $\binom{b}{a} \equiv \prod_i \binom{b_i}{a_i} \bmod p$. Taking $b = mp$ in (1), we have $b_0 = 0$ and $\binom{p-1}{a} \equiv (-1)^a$, hence (1) becomes

$$(2)\quad R(a+mp,c) \equiv (-1)^m \binom{(p-1)c-1}{mp} R(a,mp+c).$$

Forming $F(1,a-1,c)$, where $0 < a < p$, we obtain

(3) $\quad -a\,R(a,c) \equiv (-1)^a \binom{p-c_o-1}{a-1} R(1,a-1+c)$. Setting $a = p-1$, $c = 1+mp$ gives

(4) $\quad R(p-1,1+mp) \equiv -R(1,p-1+mp)$.

Using (1), $F(p-1,1,mp)$ is seen to give $0 \equiv -R(p-1,1+mp)$ for $m > 0$. This and (4) imply proposition (i).

If a is not a power of p, we may write $a = a' + p^s$ where s is such that $p^s < a < p^{s+1}$, and then $0 < a' < p^{s+1}$. In $F(a',p^s,c)$, the coefficient of $R(a'+p^s,c)$ is found to be

$$\binom{(p-1)p^s-1}{a'} \equiv \binom{p-2}{a'_s}\binom{p-1}{a'_{s-1}}\cdots\binom{p-1}{a'_o}.$$

Since this is non-zero mod p, it follows that $R(a,c)$ is reducible, hence (ii) is proved.

We shall write $R(h,k) \sim 0$ to mean that $R(h,k)$ is reducible to a multiple of $R(1,h-1+k)$. Then, by (2) and (3) we have

(5) $\quad R(h,k) \sim 0$ whenever $h_o \neq 0$.

The binomial on the right of (3) is zero whenever $a + c_o > p$. When $a + c_o = p$ we have $R(1,a-1+c) = R(1,mp-1) \equiv 0$ when $c > p$ by (i). Therefore (2) and (3) imply

(6) $\quad R(h,k) \equiv 0$ if $h_o + k_o > p$ or if $h_o + k_o = p$ and $k > p$.

6.3. <u>Lemma</u>. Any reducibility formula deduced from 5.1 remains valid if each $R(h,k)$ in the formula is replaced by $R(ph,pk)$.

<u>Proof</u>. It suffices to prove this of $F(a,b,c)$ itself. We will show that this operation performed on $F(a,b,c)$ gives $F(pa,pb,pc)$. The formula $F(pa,pb,pc)$ contains a sum of the form $\sum_{s=1}^{a} \alpha_s R(pa+pb-s, s+pc)$. By (6), when s is not of the form pt, we have $R(pa+pb-s, s+pc) \equiv 0$, hence the sum reduces to $\sum_{t=1}^{[a/p]} \alpha_{pt} R(p(a+b-t),p(t+c))$. Thus the remaining terms of $F(pa,pb,pc)$ are in $1-1$ correspondence with those of $F(a,b,c)$ under the operation. It remains to show that corresponding coefficients are equal. For the first, we have

$$\binom{(p-1)pc-1}{pb} = \binom{(p-1) + ((p-1)c-1)p}{0 + bp} \equiv \binom{(p-1)c-1}{b}.$$

The same calculation with (b,c) replaced by $(a-t,b-t)$, (t,c), and (a,b) in turn shows that the other corresponding coefficients are equal. □

The statement (iii) is a consequence of 6.1.

$F(p,b,c)$ reduces to

(7) $$(-1)^b \binom{(p-1)c-1}{b} R(p,b+c) \equiv (b_1-b_0+1)\, R(p+b,c) + (c+1)\, R(p-1+b,1+c).$$

If the p-adic expansion of k has at least two non-zero terms, we may write $k = b + p^s$ where $p^s < k < p^{s+1}$ and $b < p^{s+1}$. Taking $c = p^s$ in (7) we argue, as in the proof of (ii), that the coefficient of $R(p,b+c)$ is not zero. If also $b_0 \neq 0, 1$, then (5) implies $R(p+b,p^s) \sim 0$ and $R(p-1 + b, 1 + p^s) \sim 0$. Therefore (7) yields

(8) $R(p,k) \sim 0$ if $k > p$ and $k_0 \neq 0, 1$.

If, in (7), we take $c = 2$ and assume that $2 \leq b < p$, we obtain

$$(-1)^b \binom{p-3}{b} R(p,b+2) \equiv (1-b)\, R(p+b,2) + 3\, R(p-1+b,3)$$

By (5) the terms on the right are ~ 0, therefore

(9) $R(p,k) \sim 0$ if $4 \leqq k < p$.

We claim also that

(10) $R(p,3) \sim 0$ if $p > 3$.

To prove this start with the doubly inadmissible triple $[a,b,c] = [p,2,1]$. Then $F(p,2,1)$ is not defined because $a + b > pc$ in this case. However we may apply the method of proof of 5.1 by reducing $[p,2,1]$ to normal form in two different ways and form the difference; this yields $3R(p,3) \equiv -2R(1+p,2)$, and then (5) applies to give (10).

By (8), (9), and (10) we have that $R(p,b)$ is reducible if $b > 2$ and $b_0 \neq 0,1$. The only cases claimed to be reducible in (iv) with $b_0 = 0$ are those of the form $R(p,(mp-1)p)$, and these follow from (i) by applying 6.3. We consider now those cases with $b_0 = 1$. Given b with $b_0 = 1$, define s to be the least integer such that

(11) $b = 1 + \sum_{i=1}^{s-1} (p-1)p^i + mp^s$ where $m_0 \neq p-1$.

If m were zero, then $b = 1 + p^s - p$, and (iv) does not claim that $R(p,b)$ is reducible, so we shall assume that $m \neq 0$.

Consider first the case $s = 1$. Since $b = 1 + p$ is not claimed to be reducible, we assume that $m > 1$. Then $F(p,p,1+(m-1)p)$ and $F(p,1+p,(m-1)p)$ reduce, by (7), to

(12) $(m_0-1)\, R(p,1+mp) \equiv 2\, R(2p,1+(m-1)p) + 2\, R(2p-1,2+(m-1)p)$

(13 $m_0\, R(p,1+mp) \equiv R(1+2p,(m-1)p) + R(2p,1+(m-1)p)$

By (5), $R(2p-1,2+(m-1)p) \sim 0$ and $R(1+2p,(m-1)p) \sim 0$. Hence, if we double (13) and subtract (12), we obtain $(m_0+1)\ R(p,1+mp) \sim 0$. Since $m_0 \neq p-1$, this implies $R(p,1+mp) \sim 0$.

Suppose now that b is as in (11) with $s > 1$. Then $F(p,p^s,1+mp^s-p)$ and $F(p,1+p^s,mp^s-p)$ reduce, by (7), to

(14) $\quad (m_0+1)\ R(p,b) \equiv R(p+p^s,1+mp^s-p) + 2R(p-1+p^s,2+mp^s-p)$

(15) $\quad 3(m_0+1)\ R(p,b) \equiv 0\ R(1+p+p^s,mp^s-p) + R(p+p^s,1+mp^s-p)$

By (5), $R(p-1+p^s,2+mp^s-p) \sim 0$, hence subtracting (14) from (15) gives $2(m_0+1)\ R(p,b) \sim 0$. Since $m_0 \neq p-1$ and p is odd, we have $R(p,b) \sim 0$. This completes the proof of (iv).

Statement (v) becomes (iv) when $s = 1$. Assuming that (v) holds when $s = r - 1$, we may apply 6.3 to infer that (v) holds when $s = r$ for all $R(p^r,k)$ such that $k_0 = 0$. Suppose then that $k_0 \neq 0$. Using (7), we find that $F(p,p^s-p+1,k-1)$ reduces to

$$-\binom{(p-1)(k-1)-1}{1+p^s-p} R(p,k+p^s-p) \equiv -R(1+p^s,k-1) + k_0\ R(p^s,k)$$

Since $R(1+p^s,k-1) \sim 0$ by (5) and $k_0 \neq 0$, it follows that $R(p^s,k)$ is reducible. Note that the condition $a + b < pc$ required for the validity of $F(a,b,c)$ holds in this case for all relevant k's excepting $k = 1 + p^{s-1}$. To prove that $R(p^s,1+p^{s-1})$ is reducible, we apply the method of proof 5.1 to the doubly inadmissible triple $[p,p^s-p+1,p^{s-1}]$. In formula (1) of section 5 the $[a+b,c]$ term becomes $[1+p^s,p^{s-1}]$ which is admissible, so (2) is not required. The summation in (1) reduces to the one term

$$[p^s,1,p^{s-1}] = [p^s,R(1,p^{s-1})] + [p^s,1+p^{s-1}],$$

and $[p^s,1+p^{s-1}] = R(p^s,1+p^{s-1})$ plus admissible terms. This concludes the first reduction to normal form. As for the second, the term $[a,b+c]$ of (7) has coefficient zero because $b > (p-1)c-1$. The same is true for all terms of the sum except the last $p - 2$ terms; but, as shown in (9) and (10), these contribute no significant terms to the final reduction. The difference of the two reductions gives $R(p^s,1+p^{s-1}) \equiv 0$. This completes the proof of 6.2.

7. <u>A polynomial algebra on two generators over $\mathcal{A}(p)$</u>. Our example, applying the preceding theory, is, for any odd prime p, a polynomial algebra A on two generators u,v of degrees

$$\deg u = 4, \qquad \deg v = 2(p+1)$$

and the action of the generators $\beta, P^1, P^2, \ldots$ is given by

$$\begin{aligned}
\beta u &= 0, & \beta v &= 0,\\
P^1 u &= v, & P^1 v &= 2u^p,\\
P^2 u &= u^p, & P^i v &= 0, \text{ for } 2 < i < p \text{ and } i > p+1,\\
P^i u &= 0 \text{ for } i > 2, & P^p v &= \textstyle\sum_{i=0}^{(p-1)/2} \alpha_i u^{1+i(p+1)} v^{p-1-2i},
\end{aligned}$$

$$P^{p+1} v = v^p$$

where

(1) $$\alpha_n = \frac{1}{n+1}\binom{2n}{n}.$$

It is shown below that α_n is an integer. Since the instability conditions are satisfied for the generators, it follows that A is an unstable algebra over the free algebra $\bar{\mathcal{A}}(p)$.

According to section 4, to show that A is an $\mathcal{A}(p)$-algebra we have only to show that $R_{ij}x = 0$ for each generator x where $i < pj$ and $2j < \deg x$. For $x = u$ this means that $j = 1$ and $i = 1,2,\ldots p-1$. For $x = v$, this means $j = 1,2,\ldots p$ and $i = 1,2,\ldots pj-1$. If a particular $R_{i,j}$ is reducible in the sense of 6.1, it is a sum of terms in $\bar{\mathcal{A}}(p)$ each of which is an $R_{k\ell}$ or a composition $P^s R_{k,\ell}$ or $R_{k,\ell} P^s$. If all $R_{k,\ell}$ appearing in this sum act trivially in A, it follows that R_{ij} acts trivially. Thus we need verify only that the irreducible R_{ij} act trivially. For u there is just $R_{1,1}$, and, for v, these are $R_{1,1}, R_{1,2}, \ldots, R_{1,p}, R_{p,2}$ and $R_{p,p}$. Recalling that $R_{1,j} = P^1P^j - (j+1)P^{j+1}$ we have

$$\begin{aligned}
R_{1,1}u &= P^1P^1u - 2P^2u = P^1v - 2u^p = 0\\
R_{1,1}v &= P^1P^1v - 2P^2v = P^1u^p - 0 = 0 \text{ by } 3.3,\\
R_{1,j}v &= P^1P^jv - (j+1)P^{j+1}v = P^1 0 - 0 \text{ for } 2 \leqq j < p.
\end{aligned}$$

It remains to show that $R_{1,p}$, $R_{p,2}$ and $R_{p,p}$ are zero on v.

For this purpose we need to develop properties of the sequence of α's. The function

(2) $$g(x) = (1 - (1-4x)^{1/2})/2x$$

is readily seen to have the series expansion $\sum_0^\infty \alpha_n x^n$.

A little algebra shows that $g(x)^2 = (g(x) - 1)/x$, and this implies

(3) $$\alpha_{n+1} = \sum_{i=o}^{n} \alpha_i \alpha_{n-i}$$

Since $\alpha_o = 1$ by (1), it follows from (3) that each α_k is an integer. From (1) we obtain directly the recursion formula

(4) $$(n+1)\ \alpha_n = 2(2n-1)\ \alpha_{n-1}.$$

For a fixed prime p, we are interested only in the mod p value of α_n for $0 \leq n \leq (p-1)/2$. Consider then the sequence

(5) $$\beta_n = (-1)^n\ 2^{2n+1} \binom{(p+1)/2}{n+1} \quad \text{for } n = 0,1,2,...$$

These are integers and $\beta_o = p + 1 \equiv \alpha_o \bmod p$.

Moreover β_n satisfies the recursion formula

$$(n+1)\ \beta_n = -\ 2\ (p+1-2n)\ \beta_{n-1}$$

Subtracting this from (4) gives

$$(n+1)\ (\alpha_n - \beta_n) = 2\ (2n-1)\ (\alpha_{n-1} - \beta_{n-1}) + 2\ p\ \beta_{n-1}$$

It follows now by recursion that

(6) $$\alpha_n \equiv \beta_n \bmod p \text{ for } 0 \leq n < p - 1.$$

To show that $R_{1,p}v = 0$, we must show that P^1 applied to the sum for $P^p v$ gives v^p.

$$P^1 P^p v = \sum_n \alpha_n\ P^1\ (u^{1+n(p+1)} v^{p-1-2n})$$

$$= \sum \alpha_n\ [(1+n(p+1))\ u^{n(p+1)}\ v^{p-2n} + (p-1-2n)\ 2\ u^{(n+1)(p+1)} v^{p-2(n+1)}$$

When $n = 0$, we do have a term $\alpha_o\ v^p = v^p$. It remains to show that the other terms cancel. The conditions for this are

$$(np+n+1)\ \alpha_n + 2(p+1-2n)\ \alpha_{n-1} \equiv 0 \bmod p.$$

These follow immediately from (4) above. Next

$$R_{p,2}v = P^p P^2\ v + P^{p+2}\ v - P^{p+1}\ P^1\ v,$$

$$= P^p\ 0 + 0 - P^{p+1}\ 2u^p.$$

By 3.3, $P^{p+1}\ u^p = 0$, hence $R_{p,2}v = 0$. Finally

$$R_{p,p}\ v = P^p\ P^p\ v - 2\ P^{2p}\ v - P^{2p-1}\ P^1\ v.$$

Now $P^{2p}\ v = 0$ by definition, and $P^{2p-1}\ P^1\ v = P^{2p-1}\ 2\ u^p = 0$ by 3.3. So it is left to show that P^p applied to the sum defining $P^p\ v$ gives zero.

Consider P^p applied to a single term $u^{1+n(p+1)} v^{p-1-2n}$. According to the Cartan formula, we must form a sum over all partitions of p into a sum of $1 + n(p+1) + p - 1 - 2n = s$ non-negative terms; the term corresponding to a partition $i_1,\dots,i_s$ is the product of P^{i_k} applied to the kth factor for $k = 1,\dots,s$. Since $P^i u = 0$ for $i \neq 0,1,2$ and $P^i v = 0$ for $i \neq 0,1,p,p+1$, relatively few partitions give non-zero products. Consider first the trivial partition where all $i_k = 0$ except for one that is p. Since $P^p u = 0$, we need take of these only the $p - 1 - 2n$ partitions in which P^p acts on a v. These give

(7) $\quad (p-1-2n)\, u^{1+n(p+1)} v^{p-2-2n} \sum_j \alpha_j u^{1+j(p+1)} v^{p-1-2j}.$

Consider next a partition with k factors P^1u, ℓ factors $P^2 u$, and $p - k - 2\ell$ factors $P^1 v$. The number of such partitions is

$$\binom{1 + n(p+1)}{k,\ell}\binom{p-1-2n}{p-k-2\ell}.$$

We seek for cases (k,ℓ) where this product of binomials is non-zero mod p. Clearly $(k,\ell) = (p,0)$ is one such, and the contribution of this case is

(8) $\quad n\, u^{1+np+n-p} v^{2p-1-2n}.$

Assuming $k < p$, the denominator of the left binomial receives factors p only from $(np+n+1-(k+\ell))!$, and the numerator from $(np+n+1)!$. If $k + \ell > n + 1$, the numerator has more, and the binomial is zero mod p. Suppose then that $k + \ell \leq n + 1$. The second binomial is clearly zero unless $k + 2\ell \geq 2n + 1$. These two inequalities are satisfied only in the cases $(k,\ell) = (0,n+1)$ and $(1,n)$. Their contributions are, respectively,

(9) $\quad (p-2n-1)\, 2^{p-2-2n} u^{p(p-1)} v, \quad (n+1)\, 2^{p-1-2n} u^{p(p-1)} v$

These add mod p to $2^{p-2-2n} u^{p(p-1)} v$.

We must now multiply the sum of (7), (8), and (9) by α_n and sum for $n = 0,1,\dots(p-1)/2$. The coefficient of $u^{p(p-1)} v$ in the resulting sum is

(10) $\quad \sum_{n=o}^{(p-1)/2} 2^{p-2-2n} \alpha_n + 2\, \alpha_{(p-3)/2}\, \alpha_{(p-1)/2}\,,$

and, for $0 \leq r \leq p-2$, the coefficient of $u^{2+r(p+1)} v^{2p-3-2r}$ is

(11) $\quad (r+1)\, \alpha_{r+1} - \sum_{n=o}^{r} (2n+1)\, \alpha_n\, \alpha_{r-n}\,.$

As these comprise all terms, we have only to show that (10) and (11) are zero mod p. The latter is easy, because

$$(2n+1)\,\alpha_n\,\alpha_{r-n} + (2(r-n)+1)\,\alpha_{r-n}\,\alpha_n \;=\; (r+1)(\alpha_n\,\alpha_{r-n} + \alpha_{r-n}\,\alpha_n).$$

Using this, (11) reduces to (3). According to (6) we may replace α_n by β_n in (10). The last term of (10) is readily computed, using (5), and is found to be $-\,2^{2(p-1)}\,(p+1) \equiv -1$. The sum in (10) reduces to

$$\sum_{n=o}^{(p-1)/2} (-1)^n\, 2^{p-1} \binom{(p+1)/2}{n+1}.$$

Since $2^{p-1} \equiv 1$ mod p, and $(1+(-1))^{(p+1)/2} = 0$, this sum equals 1. Thus (10) reduces to zero. □

Remark, It can be shown with a bit more effort that any unstable polynomial algebra over $\mathcal{A}(p)$ with two generators in degrees 4 and 2(p+1) is isomorphic to the one constructed above. The case p = 3 is in fact $H^*(B\,Sp\,(2);\,Z_p)$.

Bibliography

[1] J. Adem, Relations on Steenrod powers of cohomology classes, Symposium on Algebraic Geometry and Topology, Princeton University Press, 1957.

[2] N. Steenrod, The cohomology algebra of a space, Enseignement Math. (2) 7 (1961) 153-178.

[3] T. Sugawara and H. Toda, Squaring operations on truncated polynomial algebras, Japanese J. of Math. 38 (1969) 39-50.

[4] Emery Thomas, Steenrod Squares and H-spaces: II, Annals of Math. 81 (1965) 473-495.

ON QUASI-FIBRATIONS AND ORTHOGONAL BUNDLES

Peter Hilton and Joseph Roitberg*
Battelle Memorial Institute and Cornell University
State University of New York

1. INTRODUCTION

In a previous paper ([3]; see also [2]) the authors studied the problem of classifying rank 2 H-spaces up to homotopy type. The methods used in that paper involved the study of CW-complexes of the form

(1.1) $$X = S^q \cup_\alpha e^n \cup_\beta e^{n+q}$$

which are quasi-fibrations over S^n in the sense that there is a map

(1.2) $$p: X, S^q \to S^n, pt$$

which induces homotopy isomorphisms. Such quasi-fibrations may also be called (generalized) sphere-bundles over spheres. In an appendix to [3] we obtained the following theorem.

Theorem 1.1 *If* X, *given by (1.1), quasi-fibers over* S^n *with* $q = 3$ *and* $n = 5$ *or* 7, *then* X *has the homotopy type of an orthogonal* S^q*-bundle over* S^n.

The method of proof was simply by enumeration. A careful study of the self-homotopy-equivalences of $C_\alpha = S^q \cup_\alpha e^n$ leads to the conclusion that there are precisely 3 such quasi-fibrations if $n = 5$ and precisely 10 if $n = 7$. However, James and Whitehead, in their classical study of sphere-bundles over spheres, showed that there are precisely 3 homotopy types of orthogonal S^3-bundles over S^5, and Curtis and Mislin, basing themselves on the work of James and Whitehead, have recently shown that there are precisely 10 homotopy types of orthogonal S^3-bundles over S^7.

An example due to Sutherland [9] shows that this theorem does not generalize to arbitrary values of q and n. Our purpose here is to provide a whole (infinite) family of examples of S^q-quasi-fibrations over S^n which are not homotopy equivalent to orthogonal S^q-bundles over S^n. We use the methods employed in [3],

*Talk delivered by Peter Hilton at the Conference on H-spaces, Neuchâtel, August, 1970.

and referred to above, so that it is by virtue of the methods used, and not by virtue of the results obtained, that we justify presenting these results at this conference. In fact, in all our examples, we will have $q = 2$ and the first attaching map $\alpha = 0$. The latter implies the existence of a cross-section for the quasi-fibration. Thus none of our examples will be H-spaces! We also remark that our arguments (in particular, that establishing the key Lemma 2.4) do not generalize to quasi-fibrations (1.2) with $n \neq 2$, except at the cost of considerable complication of argument and conclusion.

Sutherland constructs his example so that the total space has the homotopy type of a closed, smooth manifold. He then shows that his example does not even have the homotopy type of a differentiable S^3-bundle over S^8, i.e. a fiber bundle over S^8 with fiber S^3 and structural group Diff (S^3), the group of diffeomorphisms of S^3. In fact, Sutherland observes that results from Cerf's thesis can be used to show that any differentiable S^q-bundle over S^n must be fiber homotopy equivalent to an orthogonal S^q-bundle over S^n.

Many of the examples we construct also turn out to be homotopy equivalent to closed, smooth manifolds so that the above remark applies to these examples. However, we can even go one step further and assert that none of our examples has the homotopy type of an S^2-fiber bundle over S^n, with structural group the full group of homeomorphisms of the fiber. This is due to the classical fact that the full group of homeomorphisms of S^2 has the homotopy type of the 3-dimensional orthogonal group.

One might inquire whether the manifolds we construct here have a reasonable "geometric" description. We do not attempt to give such a description for all our examples, but we do succeed in doing this for our lowest dimensional example yielding a smooth manifold, a certain S^2-quasi-fibration over S^4. To carry this out, we base ourselves on Wall's work on 6-manifolds [10].

The rest of the paper is organized as follows. In Section 2, we construct our examples and show that they have the purported properties. We also discuss the relevance of these examples to the phenomena discussed in [1]. Finally, in Section 3, we discuss in detail our 6-dimensional example. Details of this work will be found in [4].

2. CONSTRUCTION OF THE EXAMPLES

We begin by establishing notation and recalling some needed results. Let $X = X_{\alpha\beta} = S^q \cup_\alpha e^n \cup_\beta e^{n+q} = C_\alpha \cup_\beta e^{n+q}$, $2 \leq q \leq n-2$. Let $\sigma \in \pi_n(C_\alpha, S^q)$ be the generator which satisfies $\partial(\sigma) = \alpha$, let $\iota_q \in \pi_q(S^q)$ be the generator, and let $i: S^q \to C_\alpha$ and $j: C_\alpha \to (C_\alpha, S^q)$ be the respective inclusions. We shall later only be concerned with elements $\beta \in \pi_{n+q-1}(C_\alpha)$ for which $j_*\beta = [\sigma, \iota_q]$, the relative Whitehead product of σ and ι_q (see [3]). Under these circumstances, we have the following result, which is a special case of a theorem of Sasao [6].

Theorem 2.1 *If $j_*\beta = [\sigma, \iota_q]$, the natural collapsing map $C_\alpha \to S^n$ extends to a map $p: (X, S^q) \to (S^n, \text{point})$ and p is a quasi-fibration.*

As to the homotopy type of $X_{\alpha\beta}$, we have the following elementary result.

Theorem 2.2 *If $X_{\alpha\beta} \simeq X_{\alpha'\beta'}$, then $\pm\alpha' = (\pm 1) \circ \alpha$.*

Now, if E is the total space of an S^q-bundle over S^n, then certainly $E = X_{\alpha\beta}$ for some α, β. Moreover, $\alpha \in \pi_{n-1}(S^q)$ is the characteristic class of the bundle, that is, the obstruction to a cross section. Thus the bundle admits a cross section if and only if $\alpha = 0$. We thus have the following corollary to Theorem 2.2.

Corollary 2.3 *If $X = X_{\alpha\beta}$ with $\alpha = 0$ and if X has the homotopy type of the total space of an S^q-bundle over S^n, then that bundle has a cross section.*

We wish to apply Theorem 2.1 in the case $q = 2$, $\alpha = 0$. For $\alpha = 0$, $C_\alpha = S^q \vee S^n$ and we note that the condition $j_*\beta = [\sigma, \iota_q]$ is equivalent to having β of the form

$$\beta = [\iota_n, \iota_q] + \theta, \quad \theta \in \pi_{n+q-1}(S^q) \quad ; \tag{2.1}$$

we have here identified $\pi_{n+q-1}(S^q)$ with its isomorphic image $i_*(\pi_{n+q-1}(S^q)) \subset \pi_{n+q-1}(S^q \vee S^n)$. We may now state our key technical lemma.

Lemma 2.4 *Let $X_\theta = (S^2 \vee S^n) \cup_\beta e^{n+2}$ with β as in (2.1), $n \geq 3$. Then X_θ is homotopy equivalent to $X_{\theta'}$ iff $\theta = \pm\theta'$. In particular, X_θ is homotopy equivalent to $S^2 \times S^n$ iff $\theta = 0$.*

As a corollary of Lemma 2.4, we have the main theorem.

Theorem 2.5 If $\theta \neq 0$, $n \geq 4$, then X_θ is an S^2-quasi-fibration over S^n but is not homotopy equivalent to an S^2-bundle over S^n with structural group Top (S^2), the group of all self-homeomorphisms of S^2.

We now discuss circumstances in which the spaces X_θ are homotopy equivalent to closed, smooth manifolds. Obviously, the X_θ are Poincaré Duality spaces, and are 1-connected. In order to put a manifold structure on X_θ, by the Browder-Novikov Theorem, we must study the Spivak normal spherical fibration $\nu = \nu(X_\theta)$ of X_θ (cf. [8]) and determine whether ν can be lifted to a vector bundle over X_θ. Now the easiest sufficient condition for ν to come from a vector bundle is simply that ν be (fiber homotopy) trivial. This condition, in turn, is equivalent to having X_θ stably reducible, i.e. $([\iota_2, \iota_n] + \theta) \in \operatorname{Ker} \Sigma^N$, where

$$\Sigma^N: \pi_{n+1}(S^2 \vee S^n) \to \pi_{N+n+1}(S^{N+2} \vee S^{N+n})$$

is the N-fold iterated suspension, N large. As $\Sigma([\iota_2, \iota_n]) = 0$, we see that the stable reducibility of X_θ is equivalent to the stable triviality of θ. We thus conclude:

Theorem 2.6 Let $\theta \in \operatorname{Ker} \Sigma^N: \pi_{n+1}(S^2) \to \pi_{N+n+1}(S^{N+2})$, $n \geq 3$, and suppose further that $n \neq 4k$, $k \neq 1, 3$. Then X_θ has the homotopy type of a closed, smooth π-manifold of dimension $n + 2$.

Examples where the hypotheses of Theorem 2.6 are satisfied are, of course, numerous. The lowest dimensional example where Theorem 2.6 applies occurs for $n = 5$, θ any non -0 element of $\pi_6(S^2) = \mathbb{Z}_{12}$; it is well known that $\Sigma(\pi_6(S^2)) = 0$. There is one S^2-quasi-fibration over S^4 (the smallest possible dimension to which Theorem 2.5 applies) with $\theta \neq 0$, to which Theorem 2.6 does not apply, but which nevertheless has the homotopy type of a closed, smooth manifold. We state this as

Theorem 2.7 Let $X = X_\mu = (S^2 \vee S^4) \cup_{[\iota_2, \iota_4]+\mu} e^6$, μ the generator of $\pi_5(S^2) = \mathbb{Z}_2$. Then X has the homotopy type of a closed, smooth 6-manifold.

Of course, such a manifold cannot be stably parallelizable since μ stably suspends to a non -0 element of the 3-stem. More precise information concerning manifolds realizing X will be given in the next section.

To conclude this section, we point out that the quasi-fibrations $S^2 \to X_\theta \to S^n$ furnish further examples of the phenomena discussed in [1].

Theorem 2.8 X_θ and $S^2 \times S^n$ have isomorphic homotopy groups and integral cohomology rings, but are of different homotopy types if $\theta \neq 0$.

Of course, a similar statement hold for orthogonal bundles over spheres with cross section. The smallest dimension for which such an example exists is 5; namely, take the nontrivial S^3-bundle over S^2. Note that the nontrivial S^2-bundle over S^2 has a different cohomology ring from $S^2 \times S^2$ although its cohomology groups (as well as its homotopy groups) are isomorphic to those of $S^2 \times S^2$. (Recall that for 1-connected 4-manifolds, the homotopy type is completely determined by the integral cohomology ring structure; see Milnor [5].)

Remarks

1) Observe that the nontrivial S^q-bundle over $S^2 (q \geq 2)$, in particular, the 5-dimensional example above, fails to be a spin manifold--this is simply because the non -0 element of $\pi_2(\underline{BO}_{q+1})$ is detected by the second Stiefel Whitney class w_2. For 1-connected, 5-dimensional spin manifolds, it is known that the diffeomorphism class is completely determined by the second homology group; cf. Smale [7].

2) We shall show in Section 3 that our 6-dimensional examples (see Theorem 2.7) are all spin manifolds; hence 6 is the minimal dimension for examples exhibiting the phenomena of Theorem 2.8 in the class of 1-connected spin manifolds.

3) The 5-dimensional complex $X_\lambda = (S^2 \vee S^3) \cup_{[\iota_2,\iota_3]+\lambda} e^5$, λ the generator of $\pi_4(S^2) = \mathbb{Z}_2$, is of the type appearing in Theorem 2.1 except that the condition $n - q \geq 2$ is not met. It is true, nevertheless, that, the conclusion of Theorem 2.1 holds for this space, i.e. that X_λ is an S^2-quasi-fibration over S^3. It should be noted that X_λ cannot be realized by a smooth manifold. Indeed, the same reasoning as in Lemma 3.1 below would show that any such manifold would have to be a spin manifold and it would then follow, again using Smale's classification [7], that $X_\lambda \simeq S^2 \times S^3$, violating Lemma 2.4.

3. GEOMETRIC DESCRIPTION OF THE COMPLEX X_μ

In [10], Wall gives a classification of closed, 1-connected, torsion-free, 6-dimensional spin manifolds. If $M = M^6$ is a closed manifold of the homotopy type of the complex $X = X_\mu$ of Theorem 2.7, then M is certainly 1-connected and torsion-free. As we shall now show, M must also be a spin manifold, so that we may apply Wall's results in studying M. We may then prove

Lemma 3.1 The second Stiefel-Whitney class $w_2(X)$ vanishes. Hence, any manifold M having the homotopy type of X is a spin manifold.

In order to state the main result of this section, we must first recall a few facts; see [10] for details. Let C_3^3 be the group of isotopy classes of Haefliger knots, i.e. embeddings of S^3 in S^6, and let FC_3^3 be the group of isotopy classes of framed Haefliger knots, i.e. embeddings of $S^3 \times D^3$ in S^6. There is an obvious exact sequence

$$0 \to \pi_3(\underline{SO}_3) \to FC_3^3 \to C_3^3 \to 0 \quad ;$$

since, by Haefliger's work, $C_3^3 = \mathbb{Z}$, the sequence splits. Moreover, there is a preferred splitting, induced by a geometrically defined map $FC_3^3 \to \pi_3(\underline{SO}_3) = \mathbb{Z}$. Hence, by means of this splitting $FC_3^3 = \pi_3(\underline{SO}_3) \oplus C_3^3$, a framed Haefliger knot is characterized by a pair (m,n) of integers, $m \in \pi_3(\underline{SO}_3)$, $n \in C_3^3$. We may then prove, basing ourselves on [10],

Theorem 3.2 If M is a closed manifold having the homotopy type of X, then M can be obtained from S^6 by performing surgery on a framed Haefliger knot; the pair (m,n) characterizing the knot satisfies: (1) n is odd, (2) $m + 6n = 0$. Conversely, any manifold N obtained by surgering out such a knot is homotopy equivalent to X.

Remark

If we replace the condition "n is odd" by "n is even", we obtain all closed manifolds realizing $S^2 \times S^4$.

BIBLIOGRAPHY

[1] P. J. Hilton, "On the Homotopy Type of Compact Polyhedra", *Fund. Math.*, 61 (1967), 105-109.

[2] P. J. Hilton and J. Roitberg, "On the Classification of Torsion-Free H-Spaces of Rank 2", *Symposium on the Steenrod Algebra and its Applications, Battelle-Columbus, 1970,* Springer Lecture Notes (1971).

[3] P. J. Hilton and J. Roitberg, "On the Classification Problem for H-Spaces of Rank Two", *Comm. Math. Helv.*, (1970) (to appear).

[4] P. J. Hilton and J. Roitberg, "Note on Quasi-fibrations and Fiber Bundles", *J. Math*, (to appear).

[5] J. Milnor, "On Simply Connected 4-Manifolds", *Symposium Internacional de Topologia Algebraica, Mexico,* (1958), 122-128

[6] S. Sasao, "On Homotopy Groups of Certain Complexes", *Jour. Fac. Sci.*, University of Tokyo, Sec. 1, 8, Part 3, (1960), 605-630.

[7] S. Smale, "On the Structure of 5-Manifolds", *Ann. of Math.*, 75, (1962), 38-46.

[8] M. Spivak, "Spaces Satisfying Poincaré Duality", *Topology*, 6, (1967), 77-101.

[9] W. A. Sutherland, "Homotopy-Smooth Sphere Fibrings", *Bol. Soc. Mat. Mex.*, (196 (1966), 73-79.

[10] C. T. C. Wall, "Classification Problems in Differential Topology V", *Invent. Math.*, 1, (1966), 355-374.

SOME REMARKS ON VECTOR FIELDS

Emery Thomas
University of California

My talk will cover three related topics. First, I will review briefly recent work on the existence of vector 2-fields on manifolds; then I will state two new results giving the existence of 2-plane fields; and finally I will discuss a stable approach to vector field problems. The motivation for this work stems from the fundamental result of H. Hopf on vector fields, and I present the material in this conference with that thought in mind.

Let M^n denote an n-dimensional closed, smooth manifold, which we assume is <u>oriented</u>. By a k-field on M, k a positive integer, we mean k linearly independent tangent vector fields. The problem then is, given M determine the largest k for which M has a k-field. One hopes to express the solution in terms of numerical and algebraic invariants of M, the prototype for this being the Theorem of Hopf, which states that M has a 1-field if and only if the Euler characteristic of M, χM, vanishes. Some forty years after the appearance of Hopf's result, the solution of the 2-field problem has now been given, at least for oriented manifolds. To state these results we need a new invariant, the real semi-characteristic, $\chi_R M$, defined by

$$\chi_R M = \left(\sum_i b_{2i} \right) \bmod 2 ,$$

where b_j denotes the jth real Betti number of M. Thus, $\chi_R M \in Z_2$. In contrast, recall that

$$\chi M = \sum_i (-1)^i \, b_i \in Z .$$

Research supported by the National Science Foundation.

Finally, let $W_iM \in H^i(M;Z_2)$ denote the ith Stiefel-Whitney class of M, and σM the signature of M[6], when $\dim M \equiv 0 \mod 4$. We now can state the results.

dim M(= n)	Necessary and sufficient conditions for M to have a 2-field
$n \equiv 1 \mod 4$	$W_{n-1}M = 0$, $\chi_R M = 0$
$n \equiv 2 \mod 4$	$\chi M = 0$
$n \equiv 3 \mod 4$	M always has a 2-field
$n \equiv 4 \mod 4$, $n > 4$	$\chi M = 0$, $\sigma M \equiv 0 \mod 4$

The results for $n \equiv 2$ and $3 \mod 4$ were given in 1967 [7], as was the result for $n \equiv 1$ with the additional assumption that $W_2M = 0$ [8]. Recently, M. Atiyah has developed a method for studying vector fields which gives all the results stated in the table. D. Frank, independently, obtained the case $n \equiv 4 \mod 4$.

A natural extension of the notion of a k-field is that of a field of oriented tangent k-planes on a manifold. Such a field may be regarded simply as an oriented k-plane sub-bundle of the tangent bundle of M, classically called a k-distribution. We now consider the problem of determining when a manifold has a 2-plane field. Notice that (isomorphism classes of) oriented 2-plane bundles over M are in 1-1 correspondence with $H^2(M,\mathbb{Z})$, since $BSO(2) = K(Z,2)$. The correspondence is given by sending a 2-plane bundle ξ to its Euler class $\chi(\xi)$. Given $u \in H^2(M;2)$ we introduce a new invariant $\theta_i(u) \in H^i(M;Z_2)$. For $\varepsilon = 0, 1$, and $k \geq 0$, define

$$\theta_{2k+\varepsilon}(u) = \sum_{i+j=k} W_{2i+\varepsilon} M \cup u^j .$$

Combining recent work of Atiyah and Dupont with earlier work of the author [9], [10] we obtain the following results.

Theorem 1. Let M^n be a manifold as above with $n \equiv 1 \mod 4$, $n > 1$. Then M has a field of oriented 2-planes if and only if:

(i) there is a class $u \in H^2(M;Z)$ such that $\theta_{n-1}(u) = 0$,

(ii) $\chi_R M = 0$.

Condition (i) is the necessary and sufficient condition for M to have a 2-plane field with finite singularities, and with Euler class u . (See [9]). The obstruction to extending such a 2-plane field to all of M lies in $H^n(M;\pi_{n-1}(V_{n,2})) \approx Z_2$. Atiyah and Dupont have recently shown (see [2]) that for <u>any</u> 2-plane field with finite singularities, this top obstruction is given by the mod 2 integer $\chi_R M$, and hence condition (ii).

Theorem 2. Let M be a $4k$-manifold as above, with $k > 1$. Then M has a field of oriented 2-planes if and only if:

(i) there are classes $u \in H^2(M;Z)$ and $v \in H^{4k-2}(M;Z)$ such that

(a) $v \bmod 2 = \theta_{4k-2}(u)$,

(b) $(u \cup v)[M] = \chi M$;

(ii) $\chi M \equiv (-1)^k \sigma M \bmod 4$.

The proof here goes as follows. We rephrase the problem slightly by asking, for which classes $u \in H^2(M;Z)$ is there a 2-distribution ξ_u on M with $\chi(\xi_u) = u$? We showed in [9] that for all classes u there is such a 2-distribution defined on M except at a finite set of singular points. The obstruction to extending ξ_u to all of M lies in a group $Z \oplus Z_2$. We showed in [9] that the set of integers given by the Z-component of all the obstructions obtained in this way is precisely the set

$$\chi M - (u \cdot v)[M] ,$$

where v runs over all classes in $H^{4k-2}(M;Z)$ such that

$$v \bmod 2 = \theta_{4k-2}(u) .$$

(Note that by [9], $\theta_{4k-2}(u)$ always is the mod 2 reduction of an integral class.) Thus we obtain (i)(a) and (b) in Theorem 2. On the other hand Atiyah and Dupont have shown that the Z_2-component of the obstruction for any 2-plane field with finite singularities is precisely

$$\frac{\chi M - (-1)^k \sigma M}{2} \bmod 2 ,$$

which gives (ii) in the theorem.

Condition (ii) in the Theorem can be rephrased slightly, since Atiyah has shown [1] that if M has a field of oriented 2-planes, then $\chi M \equiv 0 \mod 2$. Thus we can replace (ii) by:

(ii)' $\quad \chi M \equiv 0 \mod 2 , \qquad \chi M \equiv \sigma M \mod 4$.

A solution to the 2-plane field problem on manifolds with dimension $\equiv 3 \mod 4$ has already been given [10], as well as a partial result [9] when $\dim M \equiv 2 \mod 4$.

Motivation for the study of k-plane fields comes from the fact that a manifold has a k-dimensional foliation if and only if it has a k-distribution which is completely integrable. Recent work of Haefliger [4] has reduced the study of foliations on open manifolds to a problem in homotopy theory, via a classifying space $B\Gamma_q$ for foliations of codimension q . Specifically, let N be an open n-manifold with a field of (n-q) planes ξ and let η be the normal q-plane bundle. We regard η as a map $N \to BO(q)$. There is a natural map $\pi_q : B\Gamma_q \to BO(q)$, and Haefliger shows that ξ is homotopic to a foliation if and only if η lifts to $B\Gamma_q$. Moreover, Haefliger and Milnor [4] have shown that π_q is (q+1)-connected.

Suppose now that M is a closed manifold and set $M_0 = M - pt$. We then have

Theorem 3. For each class $u \in H^2(M;Z)$ there is a foliation $F(u)$ on M_0 of dimension 2 whose Euler class is u .

The key problem, of course, is to determine which of the foliations $F(u)$ extend to foliations on all of M .

We turn now to a discussion of the notion of stability for vector fields. Let X be a complex and ξ a q-plane bundle over X , thought of as a map $X \to BO(q)$. We define the stable bundle determined by ξ to be the composite map

$$X \xrightarrow{\xi} BO(q) \longrightarrow BO \ ;$$

we write this as $[\xi]$. Recall that $[\xi]$ has geometric dimension $\leq r$ if $[\xi]$ factors through $BO(r)$.

<u>Definition</u>. An n-manifold M has a stable k-field if $[\tau_M]$ has geometric dimension $\leq n-k$. Similarly, if η is a k plane bundle over M , then η gives a stable k-plane field if $[\tau_M]-[\eta]$ has geometric dimension $\leq n-k$.

Clearly, if M has a k-field, then M has a stable k-field. When does the converse hold? We have proved [11], [12]:

Theorem 4. Let M be an oriented n-manifold, with $n \equiv 3 \bmod 4$. Suppose that M has a k-field where $k > 0$ if n even and $k > 1$ if $n \equiv 1 \bmod 4$. Let ℓ be an integer with $\ell > k$. Then M has an ℓ-field if and only if M has a stable ℓ-field.

For $m \equiv 3 \bmod 4$ we have only a partial result [11]. Define span S^n to be the maximal number of linearly independent vector fields on S^n .

Theorem 5. Let M be an n-manifold with $n \equiv 3 \bmod 4$, and let k be a positive integer with $k \leq \operatorname{span} S^n$. Then M has an k-field if and only if M has a stable k-field.

For even manifolds, similar results can be given for plane-fields.

To show the advantage of considering stable fields, we introduce the notion of a stable obstruction. Let $p : E \to BSO$ be any fibration in a Postnikov resolution of some fibration over BSO. A class $\omega \in H^*(E;G)$ will then be called a stable obstruction. Given a manifold M, we define

$$\omega[M] = \cup f^*\omega \subset H^*(M;G) ,$$

where the union is over all maps $f : M \to E$ such that $pf = [\tau]$. Suppose now that M and N are n-manifolds such that $\omega[M]$, $\omega[N]$ are non-empty.

Claim:
$$\omega[M \# N] \subset \omega[M] + \omega[N] .$$

Here $\#$ denotes connected sum of manifolds. The result follows at once from the fact that (i) $[\tau_{S^n}] = 0$ and (ii) $[\tau_{M\#N}]$ is given by the following map:

$$M \# N \longrightarrow M \vee N \xrightarrow{\tau_M \vee \tau_N} BSO \vee BSO \longrightarrow BSO .$$

For applications of this result, see Heaps [5] and D. Frank [3].

References

[1] M. Atiyah, Vector fields on manifolds, Arbeits. für forschung des landes Nordrhein-Westfalen, No. 200, 1969.

[2] ________ and J. Dupont, to appear.

[3] D. Frank, to appear.

[4] A. Haefliger, Feuilletages Sur les variétés ouvertes, Topology, 9(1970), 183-194.

[5] T. Heaps, Almost complex structures ..., <u>ibid</u>., 111-120.

[6] F. Hirzebruch, Topological methods in algebraic geometry, 3rd edition, Springer-Verlag, New York, 1966.

[7] E. Thomas, Postnikov invariants and higher order cohomology operations, Ann. Math., 85 (1967), 184-217.

[8] ________, The index of a tangent 2-field, Comment. Math. Helv., (1967), 86-110.

[9] ________, Fields of tangent 2-planes on even-dimensional manifolds, Ann. Math., 86(1967), 349-361.

[10] ________, Fields of tangent k-planes on manifolds, Invent. Math. 3(1967), 334-347.

[11] ________, Cross-sections of stably equivalent vector bundles, Quart. J. Math., 2(17), 1966, 53-57.

[12] ________, Vector fields on low dimensional manifolds, Math. Zeit., 103 (1968), 85-93.

A FINITE COMPLEX WHOSE RATIONAL HOMOTOPY IS NOT FINITELY GENERATED

Jean-Michel Lemaire
Université de Paris

The title example is obtained as a by-product of a study of the Hopf algebra $H_*(\Omega Z, Q)$ where Z is the mapping cone of a map $f: \Sigma X_2 \to \Sigma X_1$ between the suspensions of connected spaces X_1, X_2. In some sense, this study generalizes the well-known result of Bott and Samelson [1] on $H_*(\Omega \Sigma X)$, of which we make repeated use. Most results have been announced in [3] and [4], in which no details could take place. Here, again, we shall omit technicalities when they would oversize this report ; one may expect to find them in [5].

1. Some properties of graded Lie algebras

Let k be any field of characteristic zero, $\underline{H}$ the category of graded, connected, cocommutative k-Hopf algebras, $\underline{L}$ the category of graded reduced k-Lie algebras ; Milnor and Moore [6] showed that the categories $\underline{H}$ and $\underline{L}$ are equivalent. The equivalence is given by the functors $P : \underline{H} \to \underline{L}$ (primitive elements) and $U : \underline{L} \to \underline{H}$ (enveloping algebra). Moreover, $\underline{H}$ is the category of groups over the connected cocommutative graded algebras. As a matter of fact, $\underline{H}$ (or $\underline{L}$) shares many nice properties with the category of groups (over sets), and we would like to indicate the following:

(1.1) Existence of enough free objects : for every A in $\underline{H}$, there exists a free object T and an epimorphism $p : T \to A$ in the category $\underline{H}$. Here free objects are tensor algebras with the cocommutative diagonal which is trivial on the generators (i.e. the generators are primitive), and epimorphisms are surjective maps. We obtain such a pair (T,p) by choosing a section $s : QA \to \overline{A}$ of the canonical surjection ; then the associated morphism $\tilde{s}: T(QA) \to A$

is epimorphic.(we denote $T(M)$ the free Hopf algebra over the graded vector space M).

(1.2) Presentations : a presentation of an object A of $\underline{H}$ is a coexact sequence:

$$T(M_2) \xrightarrow{j} T(M_1) \xrightarrow{p} A \longrightarrow k$$

where $T(M_1)$, $T(M_2)$ are free over the graded vector spaces M_1, M_2 . This amounts to saying that p is a cokernel of j , which in turn means that A is isomorphic to the quotient of $T(M_1)$ by the (Hopf) ideal generated by $j(M_2)$.

We may choose $M_1 = QA = \mathrm{Tor}_1^A(k,k)$ and $p = \tilde{s}$ as above. Then if we define $I = \tilde{s}^{-1}(0)$ and $QI = I/(I.QA + QA.I)$, any section $QI \longrightarrow I$ of the canonical surjection gives rise to a morphism : $T(QI) \longrightarrow T(QA)$ such that the sequence :

$$T(QI) \longrightarrow T(QA) \longrightarrow A \longrightarrow k$$

is coexact. It is not hard to exhibit an isomorphism of vector spaces $QI \xrightarrow{\sim} \mathrm{Tor}_2^A(k,k)$. Such a presentation will be called a reduced presentation. As a consequence of the above construction, a morphism in the category $\underline{H}$ is an isomorphism iff it induces an isomorphism on the Tor_1's and the Tor_2's . From this again we deduce the following analogue of the Schreier theorem:

(1.3) If T is a free object and $j: A \longrightarrow T$ is a monomorphism in $\underline{H}$, then A is a free object.

Proof: It is good enough to prove $\mathrm{Tor}_2^A(k,k) = 0$. Since T is a free A-module (cf. Milnor-Moore [6]prop.4.7) one has :

$\mathrm{Tor}_2^A(k,k) = \mathrm{Tor}_2^T(k\otimes_A T,\ k)$ and the latter is 0 because a T-projective resolution of k is given by the exact sequence :

$0 \longrightarrow \overline{T} \longrightarrow T \xrightarrow{\varepsilon} k \longrightarrow 0$. Note that the assumption that Hopf algebras are connected is essential.

(1.4) There are notions of coproduct (free product) and pushout (amalgamated sum) in the category $\underline{H}$. One may construct such objects using presentations, but we shall not insist on this point here (see for instance P.M.Cohn [2]).

Given a pushout diagram in $\underline{\mathbf{H}}$

$$\begin{array}{ccccc} & & A' & & \\ & \overset{j'}{\nearrow} & & \overset{h'}{\searrow} & \\ A & & & & B \\ & \underset{j''}{\searrow} & & \underset{h''}{\nearrow} & \\ & & A'' & & \end{array}$$

where j' and j" are monomorphisms , there is an exact sequence of the Mayer-Vietoris type :

$$\cdots \to \mathrm{Tor}_p^A(k,k) \xrightarrow[(j'_*,j''_*)]{} \mathrm{Tor}_p^{A'} \oplus \mathrm{Tor}_p^{A''} \xrightarrow[h'_*-h''_*]{} \mathrm{Tor}_p^B \xrightarrow{\partial} \mathrm{Tor}_{p-1}^A \to \cdots$$

which is similar to the one which appears in the group case (see [4]). and [9]

2. Application to rational homotopy theory

We consider the rational homotopy category $\mathcal{C}_1^{(o)}$ of (finite) 1-connected pointed CW-complexes .

Let X be an object of $\mathcal{C}_1^{(o)}$; we have the following basic facts:

(2.1) The Hurewicz map induces an isomorphism of Lie algebras (Milnor Moore [6], appendix)

$$\pi_*(\Omega X)\otimes k \xrightarrow{\sim} PH_*(\Omega X,k)$$

(2.2) The functor $H_*(\Omega ?\ , k) : \mathcal{C}_1^{(o)} \longrightarrow \underline{H}$ takes suspensions into free objects .

Indeed we have $H_*(\Omega\Sigma X) = T\widetilde{H}_*(X)$ by the Bott-Samelson theorem .

Now we may look at coexact sequences in the category $\mathcal{C}_1^{(o)}$. These are cofibre sequences . It is not true in general that $H_*(\Omega ?)$ takes coexact sequences into coexact sequences . However , we have the following result :

(2.3) Let $\Sigma X_2 \xrightarrow{f} \Sigma X_1 \xrightarrow{i} Z$ be a cofibre sequence in $\mathcal{C}_1^{(o)}$. Then : $H_*(\Omega\Sigma X_2) \xrightarrow[H_*(\Omega f)]{} H_*(\Omega\Sigma X_1) \xrightarrow[H_*(\Omega_i)]{} H_*(\Omega Z)$ is a coexact sequence in $\underline{H}$.

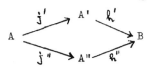

We may assume that Z is the mapping cone of f . Let $\hat{f} : X_2 \longrightarrow \Omega\Sigma X_1$ be the adjoint of $f : \Sigma X_2 \longrightarrow \Sigma X_1$. Then $\hat{f}_* : H_*(X_2) \longrightarrow H_*(\Omega\Sigma X_1)$ is the restriction of $H_*(\Omega f)$ to $H_*(X_2)$. Now $H_*(\Omega_i)$ is not epimorphic in general , therefore we do not obtain a presentation of $H_*(\Omega Z)$. Neverless if we introduce the image Λ of $H_*(\Omega_i)$ (this makes sense in <u>H</u> , cf Moore-Smith [7] §1) , the sequence :

(2.4) $H_*(\Omega\Sigma X_2) \longrightarrow H_*(\Omega\Sigma X_1) \longrightarrow \Lambda \longrightarrow k$

is a presentation of Λ , and Λ is a sub-Hopf algebra of $H_*(\Omega Z)$.

The following proposition measures the difference between Λ and $H_*(\Omega Z)$:

(2.5) <u>For all $p \geq 2$ and $q \geq 0$ there are exact sequences of vector spaces :</u>

$$0 \longrightarrow \mathrm{Tor}^{\Lambda}_{p,q}(k,k) \longrightarrow \mathrm{Tor}^{H_*(\Omega Z)}_{p,q}(k,k) \longrightarrow \mathrm{Tor}^{\Lambda}_{p+2,q-1}(k,k) \longrightarrow 0$$

<u>where the left maps are induced by the inclusion</u> $\Lambda \hookrightarrow H_*(\Omega Z)$.

<u>Moreover , such sequences hold for p = 1 if (2.4) is a reduced presentation .</u>

<u>Comments :</u> Both propositions (2.3) and (2.4) are proven using the "small" spectral sequence for the path-space fibration which arises when one filters Z by $Z_0 = \{\text{base-point}\}$, $Z_1 = \Sigma X_1$, $Z_p = Z$ for $p \geq 2$. The main steps are given in [3]. Prop. (2.3) is stated as here , but Prop.(2.5) was obtained only when the vector space $\tilde{H}_*(X_1)$ (and therefore Λ) is even-dimensional , using a degree argument . This argument can be refined in order to hold in general .

For p =1 , the assumption that (2.4) is a reduced presentation is made for the sake of simplicity : it means that f induces zero on (rational) homology , and is " as essential as possible" .

In this case , the vector space $QH_*(\Omega Z)$ is isomorphic (in a non-natural way) to the direct sum $Q\Lambda \oplus \mathrm{Tor}^{\Lambda}_3(k,k)$. Therefore $H_*(\Omega Z)$ has two kinds of generators :

- the "spherical" ones , coming from ΣX_1, which generate Λ.
- the "secondary" ones , in 1-1 correspondance with a minimal set of secondary relations in Λ .

<u>3. Sketch of the example .</u> (cf [4])

Given a presentation $T_2 \xrightarrow{j} T_1 \xrightarrow{p} \Lambda \longrightarrow k$ of a Hopf algebra Λ , we consider any map $f: \Sigma X_2 \longrightarrow \Sigma X_1$ in $\mathcal{C}_1^{(o)}$ such that :

(i) ΣX_1 and ΣX_2 are wedges of spheres .

(ii) $\tilde{H}_*(\Sigma X_i) \cong QT_i$, i = 1 or 2 .

(iii) The diagram :

$$\begin{array}{ccc} H_*(\Omega\Sigma X_2) & \xrightarrow{H_*(\Omega f)} & H_*(\Omega\Sigma X_1) \\ \downarrow \wr & & \downarrow \wr \\ T_2 & \xrightarrow{j} & T_1 \end{array}$$

, where vertical maps are Bott-Samelson isomorphisms , is commutative .

We can easily construct such a map using (2.1); by definition , its mapping cone is a CW-complex associated to the given presentation of Λ. If this presentation is reduced , we can apply Prop(2.5) In particular , if we know a finitely presented Hopf algebra whose $\mathrm{Tor}_3(k,k)$ has infinite dimension , any associated complex Z will be a finite complex such that the algebra $H_*(\Omega Z)$ is not finitely generated .

J.Stallings [9] gave a construction of a finitely presented group G with $H_3(G;\mathbb{Z})$ of infinite type . The remarks we made in §1 allow us to translate Stallings'example into the category $\underline{H}$ (or $\underline{\mathbf{L}}$) . We shall not reproduce the construction here (see [4]) but content ourselves with the result:

- Let M_1 be the graded k-vector space spanned by the elements :

  x_1 , x_2 , x_3 , x_4 of degree $p > 0$

  z of degree $q > 0$

- Let M_2 be the k-vector space spanned by the elements:

  x_{13} , x_{23} , x_{14} , x_{24} of degree $2p$

  z_1 , z_2 , z_3 of degree $p + q$

- Let $j: T(M_2) \longrightarrow T(M_1)$ defined by :

  $\forall (r,s) \in \{1,2\} \times \{3,4\}$, $j(x_{rs}) = [x_r , x_s]$

  $\forall k \in \{1,2,3\}$, $j(z_k) = [z , x_k - x_4]$

Then if Λ denotes the cokernel of j in $\underline{H}$, we obtain :

a) $\dim \operatorname{Tor}^{\Lambda}_{1,s}(k,k) = \begin{cases} 4 & \text{if } s = p \\ 1 & \text{if } s = q \\ 0 & \text{otherwise} \end{cases}$

b) $\dim \operatorname{Tor}^{\Lambda}_{2,s}(k,k) = \begin{cases} 4 & \text{if } s = 2p \\ 3 & \text{if } s = p + q \\ 0 & \text{otherwise} \end{cases}$

c) $\dim \operatorname{Tor}^{\Lambda}_{3,s}(k,k) = \begin{cases} 1 & \text{if } s = q + tp \text{ for all integers } t \geq 2 \\ 0 & \text{otherwise} \end{cases}$

d) $\dim \operatorname{Tor}^{\Lambda}_{r,s}(k,k) = 0$ if $r > 3$.

Note that a) and b) imply that the presentation :

$$T(M_2) \longrightarrow T(M_1) \longrightarrow \Lambda \longrightarrow k$$

is reduced .

Let Z be a CW-complex associated to this presentation ; Z has 5 + 7 = 12 cells and Prop.(2.5) gives :

(3.1) $\operatorname{Coker} (Q\Lambda \longrightarrow QH_*(\Omega Z)) \cong s \operatorname{Tor}^{\Lambda}_{3,*}(k,k)$ where sM stands for the suspension of the graded vector space M , i.e. $(s M)_r = M_{r-1}$;

(3.2) $\forall\, r \geq 2$, $\operatorname{Tor}^{H_*(\Omega Z)}_{r,*}(k,k) = \operatorname{Tor}^{\Lambda}_{r,*}(k,k)$.

If we choose a section $QH_*(\Omega Z) \longrightarrow \tilde{H}_*(\Omega Z)$ and a section $s \operatorname{Tor}^{\Lambda}_{3,*}(k,k) \longrightarrow QH_*(\Omega Z)$, their composite defines a morphism : $T(s \operatorname{Tor}^{\Lambda}_{3,*}(k,k)) \longrightarrow H_*(\Omega Z)$.

Now the morphism $\Lambda \amalg T(s \operatorname{Tor}^{\Lambda}_{3,*}(k,k)) \longrightarrow H_*(\Omega Z)$ which restricts to the inclusion on the first summand and to the above morphism on the second , induces an isomorphism on Tor_1 and Tor_2 and therefore is an isomorphism . This completely describes the Hopf algebra structure on $H_*(\Omega Z)$ up to isomorphism .

4 . Final remarks and questions

The method we developped proves itself successful in computing $H_*(\Omega Z, \mathbb{Q})$ for some spaces Z of catogery 2 . If cat $Z > 2$, a generalisation can be achieved in some instances , e.g. "fat wedges" of spheres in Porter's sense [8]. In this case we obtain extra generators which correspond to HOWP , and relations which generalize the Jacobi identity (mod $\mathbb{Q}$) . It would be interesting to relate extra generators to HOWP in general . (In the example , secondary generators come from $[w_k , x_1-x_2 , x_3-x_4]$ with $w_0 = z$, and for all $k \geqslant 0$, $w_{k+1} = [w_k , x_1] = \dots = [w_k , x_4]$) .

Other questions deal with the structure of finitely presented graded Lie (or Hopf) algebras over $\mathbb{Q}$:

- Is the example we constructed exceptional in nature ?
- Is the Poincaré series of a finitely presented Hopf algebra rational ? This is true for all examples we know (and for algebras with a single defining relation , by a result of Labutte's) . Note this conjecture implies that if Z is an associated complex , then $H_*(\Omega Z,\mathbb{Q})$ also has a rational Poincaré series .
- Is the Poincaré series of $\mathrm{Tor}^{\Lambda}_{r,*}(k,k)$ rational for $r > 2$ and Λ finitely presented ?

REFERENCES

[1]. R.Bott and H.Samelson , Comm.Math.Helv. 27 (1953) p.320-337.

[2]. P.M.Cohn , J.of algebra 1 (1964) p.47-69 .

[3].J.M.Lemaire , C.R.Acad.Sc.Paris 269 (1969) p.1122-1124 .

[4] " " , " " " " " p.1191-1193 .

[5] " " , Th.Sc.Math. Paris 197?

[6] J.W.Milnor and J.C.Moore , Annals of Math. 81 (1965) p.211-269.

[7] J.C.Moore and L.Smith , Amer.J.Math. 90 (1968)

[8] G.Porter , Topology 3 (1965) p.123-135.

[9] J.Stallings , Amer.J.Math. 85 (1963) p.490-503 .

Titles and references for talks given at the Conference but not included in this volume

M. ARKOWITZ	Using H-spaces to compute Whitehead products See: "Whitehead products as images of Pontryagin products". To appear in Trans. AMS.
R. BOTT	Some remarks on foliations. See: Proc. International Congress, Nice, 1970.
A. DOLD	Hopf algebras in abelian categories.
J. HUBBUCK	Polynomial algebras. To appear.
I. M. JAMES	Four properties of sphere bundles.
A. LUNDELL	Bott maps and non-stable homotopy of U(n). See: "A Bott map for non-stable homotopy of the unitary group". Topology 8 (1969) 209-217. "Bott double suspensions for U(n)". To appear.
J. C. MOORE	Cocomplete cocommutative Hopf algebras as an algebraic category.
S. WEINGRAM	On the incompressibility of certain maps. To appear in Annals of Mathematics.

H-SPACE PROBLEMS

James D. Stasheff
Temple University

Here are presented problems in the theory of H-Spaces of current interest. These problems were presented at the Conference on H-Spaces at Neuchatel (Switzerland), August 1970, and/or at the AMS Summer Colloquium in Algebraic Topology, Madison, Wisconsin, July 1970. For some problems, background material, partial or complete answers are presented.

The problems are arranged as follows:

I. Classification problems

II. Properties known for Lie groups

III. Spherical extensions

IV. H-maps and sub-H-spaces, especially torii and S^3's.

V. Homotopy properties, including infinite loop spaces

VI. Hopf algebra problems

VII. Problems about co H-Spaces and suspensions.

I. CLASSIFICATION

Our understanding of simply connected finite H-complexes would be quite satisfactory if we had a classification theorem. The appropriate notion of equivalence is H-homotopy equivalence (=H-equivalence): H-spaces (X,m) and (Y,n) are H-equivalent if there is a homotopy equivalence $h: X \to Y$ which is an H-map. For example, there are six non-H-equivalent H-space structures on S^3.

Problem 1. Classify all simply connected finite H-complexes up to H-equivalence.

Problem 2. Classify all simply connected finite H-complexes just up to homotopy equivalence. Curjel and Douglas [ ] have shown that, for a fixed dimension N, there are only finitely many such homotopy types of dimension $\leq N$. Although a given homotopy type may admit infinitely many non-H-equivalent structures, there are only finitely many group structures up to the appropriate equivalence (Curjel, Commentarii 1961). Problem 1 may thus be more reasonable for finite group complexes.

Definition: X is a finite group complex if X has the homotopy type of both a finite complex and a topological group (of course, the topological group need not be a finite complex). (Equivalently, X is an s.h.a. finite H-complex.)

Problem 3. Classify H-complexes just up to isomorphism of $H^*(X;Q)$ and of $H^*(X; Z_p)$ as Hopf algebras over the Steenrod algebra.

Problem 4. Are there finitely many homotopy types of finite H-complexes of rank n? [By rank, we always mean the number of generators of the exterior algebra $H^*(X; Q)$.] This is strictly a mod 2 problem. For $n = 2$, the answer is yes, since Hubbuck has shown that $H^*(X; Z_2) \simeq H^*(G_2; Z_2)$ if there is 2-torsion. In general one would hope for a bound on the dimension of the generators.

Problem 5. Let $A^* = \sum_{i \geq 0} A^i$ be a candidate for $H^*(X; Z)$, i.e. an associative graded simply connected Z-algebra. Let $T(A^*)$ be the set of homotopy types of simply connected finite complexes.

For which A^* is $T(A^*)$ a finite set?

Conjecture. $T(A^*)$ is finite iff $A^* \otimes Q$ is an exterior algebra on odd dimensional generators.

II. LIE PROPERTIES

Given the complete classification of simple Lie groups and extensive computations of cohomology, homotopy, etc., it is possible to observe many characteristics and then attempt to prove their existence for simply connected finite H-complexes, adding a requirement of homotopy associativity or equivalence with a loop space where obvious counter-examples such as S^7 demand it. Throughout these problems, X will denote a simply connected finite H-complex, and $X_{(p)}$ will denote X localized at p (cf. the talks of Mislin and Curtis at this conference).

Problem 6. Is $H_*(\Omega X; Z)$ torsion free? The work of Browder [1], Clark [2,3], and Gitler [7] is relevant. In particular, it is possible that $PH^2(X;Z_p) \neq QH^2(X;Z_p)$?

Problem 7. If $H_*(X; Z)$ has p-torsion, must it then have 2-torsion? The answer is "NO" in general. The counter example X due to Mislin [9] looks like $F_4 \times S^7 \times S^{19}$ mod 3 and like Sp(6) otherwise. One can easily see that X is not homotopy associative for, as usual [5], the existence of XP (3) is not compatible with the Steenrod algebra. Homotopy associativity is not enough for one can similarly use E_8 mod 5 to obtain a homotopy associative space with p-torsion only for p = 5. Thus p-torsion may imply 2-torsion for finite group complexes. May $H_*(X;Z)$ have p-torsion for $p > 5$?

Problem 8. If X is a simply connected finite group complex, $\pi_3(X) \supset Z$. Is $\pi_3(X)$ free?

Problem 9. Conjecture: $X_{(p)}$ is a finite group complex iff there is a Lie group G such that $G_{(p)} \simeq X_{(p)}$. Since the conference, this has been shown to be false, even beyond S^7 factors, for H-spaces [11].

Problem 10. Conjecture: X is a loop space iff there are Lie groups G and G' such that $X \times G' \simeq G \times G'$. The original Hilton-Roitberg example was found this way. It is conceivable that, after localizing, G' could be cancelled.

Problem 11. When is $K^*(X)$ an exterior algebra? Can generators then be chosen to be representations, corresponding to H-maps $X \to GL(n, C)$? What can be said about maps $BX \to BGL(n, C)$ if X is a finite group complex? (cf. Problem 26).

Problem 12. Does the spectral sequence from $H^*(X/T)$ collapse to $K^*(X/T)$ if T is a sub torus of X in a suitable sense (see § IV) with rank = rank X ?

Problem 13. Conjecture: $H_*(X)$ has no p-torsion iff the (generalized) Weyl group (see § IV) acts on $H^*(X/T; Z_p)$ as a permutation group. This has recently been proved by Pittie for Lie groups.

Problem 14. If $H^*(X;Z) \approx E(x_1,\ldots,x_r)$ with $\dim x_i \leq \dim x_{i+1}$, when does there exist a map $X \to S^{\dim x_r}$ so that $H_*(X:Z)$ maps onto $H_*(S;Z)$? When the map exists, when is the fiber an H-space of rank $r - 1$?

III. SPHERICAL EXTENSIONS

By spherical extensions X of a finite H-complex Y we mean fibrations of the form $Y \to X \to S^n$. In particular, we consider principal H-bundles $H \to G \xrightarrow{p} G/H = S^n$ and pull back by $S^n \to S^n$ of degree k to get $H \to E_k \to S^n$.

Problem 15. For $n > 7$ is E_k an H-space if k is odd? Zabrodsky [12] has shown k odd is necessary if $p^*(u) \neq 0$ where u generates $H^n(S^n)$.

Problem 16. [1st approximation to 15]. Does E_k have trivial Whitehead products?

Problem 17. For $S^3 \to E_k \xrightarrow{p} S^7$, do E_k and S^7 admit multiplications such that p is an H-map? The multiplication on E_k could not be homotopy associative (Stasheff).

Problem 18. Is E_k homotopy associative for $S^3 \to E_4 \to S^7$ or $SU(3) \to E_3 \to S^7$? Rector says "No" for E_4.

Problem 19. Let $S^3 \to X \to S^{2p+1}$ be classified by α_1, generating $Z_p \subset \pi_{2p}(S^3)$. Is $X_{(p)}$ a loop space for $p > 5$? For $p = 3$, $X \underset{3}{\sim} Sp(2)$ and for $p = 5$, $X \underset{5}{\sim} G_2$. [Zabrodsky has shown $S^3_{(p)}$ is not a double loop space for $p > 3$; it was already known $S^3_{(2)}$ and $S^3_{(3)}$ were not homotopy commutative.]

IV. H-maps and sub-H-spaces, especially torii and S^3's .

From a homotopy point of view, it is natural to consider H-maps or, for groups, strongly homotopy multiplicative maps. For finite complexes or even Lie groups, many basic questions remain unanswered. Recall that homotopy classes of shm maps $H \to G$ correspond to $[B_H, B_G]$, homotopy classes of maps $B_H \to B_G$.

Problem 20. For connected Lie groups, do the homotopy classes of Hom (H,G) correspond to $[B_H, B_G]$?

Problem 21. For simply connected simple Lie groups, a non-trivial homomorphism $G \to G$ is a rational homotopy equivalence. Is the same true for simply connected finite H-complexes, or for s.h.m. maps if G is a finite group complex?

Problem 22. For what degrees d is a map $S^3 \to S^3$ of degree d an s.h.m. map? Equivalently, for which d does there exist $f\colon BS^3 \to BS^3$ of degree d in dimension 4? Using K-theory, several people can show d must be zero or of the form $(2r+1)^2$. The question can also be asked in terms of the exotic multiplications on S^3 which admit classifying spaces. For H-maps, a complete solution has been given by Arkowitz and Curjel, as it has for H-maps on S^7 by Mislin.

Problem 23. Classify the strongly homotopy associative (s.h.a.) structures on S^3. The work of Slifker shows there may be infinitely many.

Problem 24. If α is a generator of infinite order of $\pi_3(X)$, then is α represented by an H-map? If X is a finite group complex, is α represented by an shm map? (The exotic structures on

S^3 should be considered as needed). Rector [10] has shown the exotic s.h.a. structures on S^3 admit sha maps into sha structures on $\Xi_{\pm 5}$, the Hilton-Roitberg examples. A candidate for a representative of α, suggested by the Lie case, is $f: S^3 \to X$ given by choosing a sub-S^1 of X and extending continuously to some $g: e^2 \to X$ and then defining f on

$S^3 = S^1 \wedge S^2$

by $f(x,y) = xg(y)x^{-1}$

$f(*,y) = g(y)$.

More specifically it is natural to look at sub-S^3's and, more classically, sub-torii of finite H-complexes. First we have three intensities for our definition of "sub".

Definition (Rector): If H, G are finite group complexes, H is a subgroup of G means a map $f: B_H \to B_G$ with fibre a finite complex, called G/H. (This gives a principal fibration $H \to G \to G/H$ with s.h.m. $H \rightarrowtail G$.)

Definition (Moore): If H is a finite group complex and X a finite H-complex, H is a subgroup of X means a finite complex 'G/H' and a principal fibration $H \to G \to$ 'G/H' such that $H \to G$ is an H-map.

Definition (Stasheff): If Y, X are finite H-complexes, Y is a sub H-space of X means a finite complex "G/H" and a map $G \to$ "G/H" with fibre of the homotopy type of H such that inclusion is equivalent to an H-map.

Using the above approach, we can talk about faithful representations of a finite group complex.

Problem 25. Describe faithful representations $X \hookrightarrow U(n)$ for finite group complexes X.

We can also look for classical subgroups of exotic H-complexes. Rector [10] focuses attention on sub-torii and sub-S^3's. Recall that there are 8 homotopy classes of multiplications on S^3 which extend to sha structures (equivalently there are at least 4 homotopy types for BS^3). Rector shows that there are finite group complex structures on S^3 which do not have S^1 as a subgroup.

Problem 26. For which multiplications on S^3, is S^1 a sub-H-space? For a Lie group G, a maximal torus has the same rank as G. Rector suggests defining maximal torus of a finite group complex to mean a sub-group $T = S^1 \times \ldots \times S^1$ of X with rank equal to that of X.

Problem 27. If X is a connected finite group complex and $T_i \xrightarrow{t_i} X \to X/T_i$ are maximal torii, is B_{t_1} homotopic to B_{t_2} ?

Keeping in mind the difficulty implied in Problem 27, Rector [10] defined a generalized Weyl group as follows:

Definition. Let $f: B_T \to B_X$ be a subtorus. The Weyl group $W(X,f) = \{\alpha \varepsilon [B_T, B_T] \mid \alpha$ is a homotopy equivalence and $f\alpha \simeq f\}$. Notice $W(X,f)$ can be regarded as a subgroup of $GL(n,Z)$, n = rank T.

Rector shows $W(G,f)$ is isomorphic to the classical Weyl group if G is a connected Lie group and f is induced by the inclusion of a Lie maximal torus T. He proposes the following problems, motivated by the classical structure.

Problem 28. For a connected finite group complex X, is $W(X,f)$ independent of f if rank T = rank X? at least if X is Lie?

Problem 29. For a finite group complex $W(X,f)$ and $f: B_T \to B_X$ a maximal torus, is $H^*(B_X;Q)$ isomorphic with the subspace of $H^*(B_T;Q)$ invariant under $W(X,f)$?

By analogy with the complex case, we can consider quaternionic torii $S^3 \times \ldots \times S^3$.

Problem 30. Study the existence of quaternionic torii as subgroups of finite group complexes. In particular if $T^r \to G$ is a maximal torus, what is the maximal k such that we have subgroups $T^k \times T^{r-k} \to (S^3)^k \times T^{r-k} \to G$?

In addition to the torii abelian subgroups of classical interest are the finite central subgroups.

Problem 31. Given a finite group complex G, when can one find a finite abelian subgroup π so that G/π is a finite group complex (i.e., when can one find a finite group complex H and a fibration $K(\pi,1) \to B_G \to B_H$?)? If π is a finite subgroup of a maximal torus of G and π is invariant under the action of $W(G,f)$, when is G/π a finite group complex?

In the category of compact simply connected groups G, if H is a normal subgroup, we have $G \equiv H \times G/H$ as topological groups. This is extremely important in reducing classification to that of simple groups.

Problem 32. If H is a subgroup of the finite group complex G such that $G \to G/H$ is an shm map, is $B_G \simeq B_H \times B_{G/H}$? or at least $G \simeq H \times G/H$ as spaces?

More generally, one can consider the problem of extensions $H \to G \to K$ whether in terms of finite complexes or not. In the category of topological groups, with the assumption that extensions are principal H-bundles, H abelian, a reasonable solution has recently become available. Heller has shown that such extensions can be described in terms of appropriate homological algebra. Independently, Dennis Johnson and Graeme Segal have defined complexes C so that $H^2(C)$ classifies such extensions. In particular, they both provide an exact sequence

$$0 \to H^2_c(K;H) \to \text{Ext}(K;H) \to [B_K, B_H]$$

where $H^*_c(K;H)$ is the cohomology of K with coefficients in H defined entirely in terms of continuous cochains $K \times \ldots \times K \to H$. The map $\text{Ext}(K;H) \to [B_K, B_H]$ classifies the bundle.

Problem 33. a) Classify extensions (suitably redefined perhaps) of s.h.a. H-spaces;

b) Classify extensions of H-spaces.

One solution to 33a) is to define equivalence so that the classification reduces to that of the fibrations $B_H \to B_G \to B_K$ which, for reasonable spaces, is given by $[B_K, \mathcal{H}(B_H)]$ when $\mathcal{H}(\ )$ denotes "homotopy equivalences of".

V. Homotopy properties, including infinite loop spaces.

For a single space X, it is natural to look at the set M(X) of equivalence classes of multiplications on X; i.e., equivalent via an H-homotopy equivalence.

Problem 34. Is every multiplication m on a finite H-complex X H-equivalent to its transpose m T where $T(x,y) = (y,x)$?

In the study of 2-stage Postnikov systems and more generally of infinite loop spaces, the following question arises:

Problem 35. Study $\Omega : M(X) \to M(\Omega X)$ given by looping the multiplication. In particular, what can be said about the kernel and image of Ω ?

Problem 36. What is the homotopy nilpotency of a connected finite H-complex? (i.e. the smallest n s/t the n-fold commutator, $G^n \to G$ by $g_1,\ldots,g_n \to [\ldots[g_1,g_2]\ldots,g_n]$, is null homotopic.) Is it finite? Is it equal to the rank? (The answer is yes for torii, S^3 and SU(3).)

Problem 37. If X is a homotopy associative H-space, mod 2 equivalent to a product of spheres, must the spheres all have dimensions 1 or 3?

Recent work of Boardman - Vogt, Beck, Segal-Anderson, May, Madsen and Tsuchiya has exhibited many infinite loop spaces including $G = F = \lim\{\text{homotopy equivalences } S^n \to S^n\}$ and quotients thereof. There are open questions as to uniqueness of the structures implied.

<u>Problem 38</u>. Compare the various iterated classifying spaces for G, Top, PL, O, U, etc. In particular how many possible infinite loop structures are there on B_U or B_O?

<u>Problem 39</u>. Describe in terms suitable for computation the above spaces and quotients F/Top, F/PL, F/O, etc.

<u>Problem 40</u>. Compute $H^*(B_F;Z_p)$ as an algebra over $\mathcal{A}(p)$. Describe generators in some meaningful way e.g. via higher order Bocksteins, Dyer-Lashof operators or as generalized Gitler - Stasheff classes.

<u>Problem 41</u>. As an infinite loop space, does G/Top mod $p > 2$ have a factor X of the homotopy type of $\Pi K(Z,4k)$?

<u>Problem 42</u>. Consider $X = \Pi K(Z_2, 2^i n)$ with multiplication m such that the duals α_{2^i} of the fundamental classes satisfy $m_*(\alpha_{2^i} \otimes \alpha_{2^i}) = \alpha_{2^{i+1}}$. Does this multiplication correspond to an infinite loop structure on X? (D. Kraines and D. S. Kahn have partial results. k-invariants of the Postnikov system should correspond to the relations $Sq^{n+1}\iota_n = 0$, $Sq^{2n+1}Sq^{n+1} = 0$, $Sq^{2^i n+1} \ldots Sq^{n+1} = 0$.

<u>Problem 43</u>. Let p be prime. Given any X, define $F_oX = X$ and proceed by induction. Given $x_n \in H^{k_n}(F_nX;Z)$ of order p^{r_n} where F_nX has no p - torsion in dimensions $m < k_n$, let $F_{n+1}X$ be the fibre of $F_nX \to K(Z,k_n)$. Let $FX = \lim F_nX$. If X is an H-space , so is FX. Is FK(Z,n) an infinite loop space?

V1. Hopf algebra problems.

Problem 44. If X is an H-space complex, can the multiplication be changed so that $H_*(X;Q)$ is associative.(Curjel has proved this if X is finite dimensional over Q.)

Problem 45. For any Y, let Y_q be the q-fold reduced product of James, so that $Y_\infty \simeq \Omega SY$. If Y is $\Sigma^m X$ and k is a field, is $\sigma_* : QH_*(\Omega^{j+1} Y_q;k) \to PH_*(\Omega^j Y_q;k)$ surjective for $j < m$. (This is easy for k of characteristic zero and follows from results of Milgram if $q = \infty$.)

Suppose that k is a field and $f : A \to B$ in a morphism of Hopf algebra.

Problem 46. Under what condition is B a free left A-module?(e.g. if f is a monomorphism and A and B are cocomplete, then this is known to be the case). In particular is it sufficient for B to be a free A-module that A and B are cocommutative with involution?

Problem 47. Under what condition is B faithfully flat as a left A-module?

Problem 48. Under what conditions is A injective as a left B comodule ? (e.g. if f is an epimorphism and A and B are cocomplete, then this is known to be the case). In particular is it sufficient for A to be an injective B-comodule that A and B are cocommutative with involution?

Problem 49. If A and B are cocommutative under what condition is f the composite of a normal epimorphism followed by a monomorphism (this is known to be the case if A and B are cocomplete).

Problem 50. If A and B are commutative under what conditions is f the composite of an epimorphism followed by a normal monomorphism (this is known to be the case if A and B are connected or more generally locally nilpotent)?

VII. H-spaces and suspensions.

Problem 51. Classify sphere fibrations over spheres up to homotopy type. (cf. James and Whitehead [8] for sphere bundles).

Problem 52. If K is a simply connected finite complex, is there a fibration $F \to E \to K$ such that E has the rational homotopy type of a suspension and $\exists N$ s/t $\pi_q(F) \otimes Q = 0$ for $q > N$?

Problem 53. Is every coH-space the wedge of a simply connected coH-space and a bouquet of circles?

Problem 54. Find an (n-1)-connected coH-space Y of dim $> 4n-5$ which is homotopy coassociative but not of the (primitive = coH-) homotopy type of a suspension. (For Y of dimension $\leq 4n-5$, Ganea [6] has shown homotopy coassociative implies being of the primitive homotopy type of a suspension.)

Problem 55. For p odd, $S^3 \cup_{\alpha_1} e^{2p+1}$ is a coH-space. Can it carry homotopy cocommutative comultiplications, or does it always have homotopy conilpotency equal to p?

References

[1] Browder, William, On differential Hopf algebras. Trans. AMS 107 (1963), 153-176.

[2] Clark, Allan, Homotopy commutativeity and the Moore spectral sequence. Pac. J. M. 15 (1965), 65-74.

[3] Clark, Allan, Hopf algebras over Dedekind domains and torsion in H-spaces. Pacific J. Math. 15 (1965), 419-426.

[4] Curjel, Caspar; Douglas, Roy, On H-spaces of finite dimension, Proc. Adv. Study Inst. Alg. Top., Aarhus, 1970 (also this conference).

[5] Douglas, Roy; Sigrist, François, Homotopy - associative H-spaces which are sphere bundles over spheres. Comm. Math. Helv. 44 (1969), 308-9.

[6] Ganea, Tudor, Cogroups and suspensions. Inv. math. 9 (1970), 185-197.

[7] Gitler, Samuel, Spaces fibred by H-spaces. Bol. Soc. Mat. Mex (2) 7 (1962), 71-84.

[8] James, Ioan; Whitehead, J.H.C., The homotopy theory of sphere bundles over spheres, I and II. Proc. LMS (3) 4 (1954), 196-218
5 (1955), 148-166.

[9] Mislin, Guido, H-spaces mod p, I., this conference.

[10] Rector, David, Subgroups of finite dimensional topological groups (preprint).

[11] Stasheff, James, Non-Lie finite H-complexes, even mod p > 3 (preprint).

[12] Zabrodsky, Alexander, this conference.

BIBLIOGRAPHY ON H-SPACES

I. M. James
University of Oxford

Instead of the text of my lecture at the conference (which will be published in the Oxford Quarterly Journal, and would seem rather out of place here), I venture to contribute a list of publications which seem to me to be particularly concerned with H-spaces. Separate bibliographies might be compiled for subjects such as the theory of Lie groups or Hopf algebras. In areas such as these I have had to be selective. But in the central region, as I see it, I have tried to list everything I know of which has been published by October 1970.

1. ADAMS, J.F. On the cobar construction. Proc. Nat. Acad. Sci. U.S.A. 42 (1956), 409-412.

2. ADAMS, J.F. On the non-existence of elements of Hopf invariant one. Bull. Amer. Math. Soc. 64 (1958), 279-282.

3. ADAMS, J.F. On the non-existence of elements of Hopf invariant one. Ann. of Math. 72 (1960), 20-104.

4. ADAMS, J.F. On Chern characters and the structure of the unitary group. Proc. Cambridge Philos. Soc. 57 (1961), 189-199.

5. ADAMS, J.F. The sphere, considered as an H-space mod p. Quart. J. Math. Oxford (2) 12 (1961), 52-60.

6. ADAMS, J.F. H-spaces with few cells. Topology 1 (1962), 67-72.

7. ADAMS, J.F. and ATIYAH, M.F. K-theory and the Hopf invariant. Quart. J. Math. Oxford (2) 17 (1966), 31-38.

8. ADAMS, J.F. and HILTON, P.J. On the chain algebra of a loop space. Comment. Math. Helv. 30 (1956), 305-330.

9. AL'BER, S.I. Homologies of the spinor group. Dokl. Akad. Nauk. SSSR 104 (1955), 341-344. (Russian)

10. AL'BER, S.I. Homologies of a non-oriented loop space and their application to the calculus of variations in the large. Dokl. Akad. Nauk. SSSR 155 (1964), 13-16. (Russian)

11. ARAKI, S. On the homology of the spinor groups. Mem. Fac. Sci. Kyusyu Univ. 9 (1955), 1-35.

12. ARAKI, S. On the non-commutativity of Pontrjagin rings mod 3 of some compact exceptional groups. Nagoya Math. J. 17 (1960), 225-260.

13. ARAKI, S. A correction to my paper "On the non-commutativity of Pontrjagin rings mod 3 of some compact exceptional groups". Nagoya Math. J. 19 (1961), 195-197.

14. ARAKI, S. Cohomology modulo 2 of the compact exceptional groups E_6 and E_7. J. Math. Osaka City Univ. 12 (1961), 43-65.

15. ARAKI, S. On Bott-Samelson K-cycles associated with symmetric spaces. J. Math. Osaka City Univ. 13 (1962), 87-133.

16. ARAKI, S. On cohomology mod p of compact exceptional Lie groups. Sugaku 14 (1962/63), 219-235. (Japanese)

17. ARAKI, S. Differential Hopf algebras and the cohomology mod 3 of the compact exceptional groups E_7 and E_8. Ann. of Math. 73 (1967), 404-436.

18. ARAKI, S. Hopf structures attached to K-theory; Hodgkin's theorem. Ann. of Math. 85 (1967), 508-525.

19. ARAKI, S., JAMES, I.M. and THOMAS, E. Homotopy-abelian Lie groups. Bull. Amer. Math. Soc. 66 (1960), 324-326.

20. ARAKI, S. and KUDO, T. Topology of H_n-spaces and H-squaring operations. Mem. Fac. Sci. Kyusyu Univ. 10 (1956), 85-120.

21. ARAKI, S. and KUDO, T. On $H_*(\Omega^N(S^n); Z_2)$. Proc. Japan Acad. 32 (1956), 333-335.

22. ARAKI, S. and SHIKATA, Y. Cohomology mod 2 of the compact exceptional group E_8. Proc. Japan Acad. 37 (1961), 619-622.

23. ARKOWITZ, M. Homotopy products for H-spaces. Michigan Math. J. 10 (1963), 1-9.

24. ARKOWITZ, M. A homological method for computing certain Whitehead products. Bull. Amer. Math. Soc. 74 (1968), 1079-1082.

25. ARKOWITZ, M. and CURJEL, C.R. On the number of multiplications of an H-space. Topology 2 (1963), 205-209.

26. ARKOWITZ, M. and CURJEL, C.R. Homotopy commutators of finite order II. Quart. J. Math. Oxford (2) 15 (1964), 316-326.

27. ARKOWITZ, M. and CURJEL, C.R. Groups of homotopy classes. Lecture Notes in Mathematics vol. 4 (Springer 1964).

28. ARKOWITZ, M. and CURJEL, C.R. On maps of H-spaces. Topology 6 (1967), 137-148.

29. ARKOWITZ, M. and CURJEL, C.R. Some properties of the exotic multiplications on the three-sphere. Quart. J. Math. Oxford (2) 20 (1969), 171-176.

30. ATIYAH, M.F. On the K-theory of compact Lie groups. Topology 4 (1965), 95-99.

31. BARCUS, W.D. and MEYER, J-P. The suspension of a loop space. Amer. J. Math. 80 (1958), 895-920.

32. BARRATT, M.G. and MAHOWALD, M.E. The metastable homotopy of O(n). Bull. Amer. Math. Soc. 70 (1964), 758-760.

33. BAUM, P.F. Cohomology of homogeneous spaces. Bull. Amer. Math. Soc. 69 (1963), 531-533.

34. BAUM, P.F. On the cohomology of homogeneous spaces. Topology 7 (1968), 15-38.

35. BAUM, P.F. and BROWDER, W. The cohomology of quotients of classical groups. Topology 3 (1965), 305-336.

36. BERSTEIN, I. and GANEA, T. Homotopical nilpotency. Illinois J. Math. 5 (1961), 99-130.

37. BOARDMAN, J.M. and VOGT, R.M. Homotopy-everything H-spaces. Bull. Amer. Math. Soc. 74 (1968), 1117-1122.

38. BOREL, A. Sur la cohomologie des variétés de Stiefel et de certains groupes de Lie. C.R. Acad. Sci. Paris 232 (1951), 1628-1630.

39. BOREL, A. La cohomologie mod 2 de certains espaces homogènes. Comment. Math. Helv. 27 (1953), 165-197.

40. BOREL, A. Commutative subgroups and torsion in compact Lie groups. Bull. Amer. Math. Soc. 66 (1960), 285-288.

41. BOREL, A. Sur la cohomologie des espaces fibrés principaux et des espaces homogènes de groupes de Lie compacts. Ann. of Math. 57 (1963), 115-207.

42. BOREL, A. Homology and cohomology of compact connected Lie groups, Proc. Nat. Acad. Sci. USA 39 (1953), 1142-1146.

43. BOREL, A. Sur l'homologie et la cohomologie des groupes de Lie compacts connexes. Amer. J. Math. 76 (1954), 273-342.

44. BOREL, A. Topology of Lie groups and characteristic classes. Bull. Amer. Math. Soc. 61 (1955), 397-432.

45. BOREL, A. Sur la torsion des groupes de Lie. J. Math. pures appl. 35 (1956), 127-139.

46. BOREL, A. Sousgroupes commutatifs et torsion des groupes de Lie compacts. Tohoku Math. J. 13 (1961), 216-240.

47. BOREL, A. and SERRE, J-P. Determination des p-puissances réduites de Steenrod dans la cohomologie des groupes classiques. Applications. C.R. Acad. Sci. Paris 233 (1951), 680-682.

48. BOREL, A. and SERRE, J-P. Groupes de Lie et puissances réduites de Steenrod. Amer. J. Math. 75 (1953), 409-448.

49. BOREL, A. and CHEVALLEY, C. The Betti numbers of the exceptional groups. Mem. Amer. Math. Soc. 14 (1955), 1-9.

50. BOREL, A. and HIRZEBRUCH, F. Characteristic classes and homogeneous spaces I. Amer. J. Math. 80 (1958), 458-538.

51. BOREL, A. and HIRZEBRUCH, F. Characteristic classes and homogeneous spaces II. Amer. J. Math. 81 (1959), 315-382.

52. BOREL, A. and HIRZEBRUCH, F. Characteristic classes and homogeneous spaces III. Amer. J. Math. 82 (1960), 491-504.

53. BOTT, R. On torsion in Lie groups. Proc. Nat. Acad. Sci. USA 40 (1954), 586-588.

54. BOTT, R. An application of the Morse theory to the topology of Lie groups. Bull. Soc. Math. France 84 (1956), 251-281.

55. BOTT, R. The stable homotopy of the classical groups. Proc. Nat. Acad. Sci. USA 43 (1957), 933-935.

56. BOTT, R. The space of loops on a Lie group. Michigan Math. J. 5 (1958), 35-61.

57. BOTT, R. The stable homotopy of the classical groups. Ann. of Math. 70 (1959), 313-337.

58. BOTT, R. Quelques remarques sur les théorèmes de périodicité. Bull. Soc. Math. France 87 (1959), 293-310.

59. BOTT, R. A note on the Samelson product in the classical groups. Comment. Math. Helv. 34 (1960), 249-256.

60. BOTT, R. An application of the Morse theory to the topology of Lie groups. Proc. Internat. Congress Math. 1958, 423-426. (Cambridge Univ. Press, 1960).

61. BOTT, R. A report on the unitary group. Proc. Symp. Pure Math. III, 1-6. (American Mathematical Society, 1961).

62. BOTT, R. and SAMELSON, H. On the Pontrjagin product in spaces of paths. Comment. Math. Helv. 27 (1953), 320-337.

63. BOTT, R. and SAMELSON, H. The cohomology ring of G/T. Proc. Nat. Acad. Sci. USA 41 (1955), 490-493.

64. BOTT, R. and SAMELSON, H. Applications of the theory of Morse to symmetric spaces. Amer. J. Math. 80 (1958), 964-1029.

65. BOTT, R. and SAMELSON, H. Applications of Morse theory to symmetric spaces. Symposium internacional de topologia algebraica, 282-284. (Universidad Nacional Autonoma de Mexico and UNESCO, Mexico City, 1958).

66. BOTT, R. and SAMELSON, H. Applications of the theory of Morse to symmetric spaces. Correction. Amer. J. Math. 83 (1961), 207-208.

67. BROWDER, W. The cohomology of covering spaces of H-spaces. Bull. Amer. Math. Soc. 65 (1959), 140-141.

68. BROWDER, W. Homology operations and loop spaces. Illinois J. Math. 4 (1960), 347-357.

69. BROWDER, W. Loop spaces of H-spaces. Bull. Amer. Math. Soc. 66 (1960), 316-319.

70. BROWDER, W. Homology and homotopy of H-spaces. Proc. Nat. Acad. Sci. USA 46 (1960), 543-545.

71. BROWDER, W. Torsion in H-spaces. Ann. of Math. 74 (1961), 24-51.

72. BROWDER, W. Fiberings of spheres and H-spaces which are rational homology spheres. Bull. Amer. Math. Soc. 68 (1962), 202-203.

73. BROWDER, W. Homotopy commutative H-spaces. Ann. of Math. 75 (1962), 283-311.

74. BROWDER, W. Higher torsion in H-spaces. Trans. Amer. Math. Soc. 108 (1963), 353-375.

75. BROWDER, W. On differential Hopf algebras. Trans. Amer. Math. Soc. 107 (1963), 153-176.

76. BROWDER, W. Homology rings of groups. Amer. J. Math. 90 (1968), 318-333.

77. BROWDER, W. and NAMIOKA, I. H-spaces with commutative homology rings. Ann. of Math. 75 (1962), 449-451.

78. BROWDER, W. and SPANIER, E. H-spaces and duality. Pacific J. Math. 12 (1962), 411-414.

79. BROWDER, W. and THOMAS, E. On the projective plane of an H-space. Illinois J. Math. 7 (1963), 492-502.

80. BROWN, R.F. H-manifolds have no non-trivial idempotents. Proc. Amer. Math. Soc. 24 (1970), 37-40.

81. BRUMFIEL, G. On the homotopy groups of BPL and PL/O. Ann. of Math. 88 (1968), 291-311.

82. CHANG, S.C. On Jacobi identity. Acta Math. Sinica 4 (1954), 365-379. (Chinese. English summary)

83. CLARK, A. On π_3 of finite dimensional H-spaces. Ann. of Math. 78 (1963), 193-196.

84. CLARK, A. Hopf algebras over Dedekind domains and torsion in H-spaces. Pacific J. Math. 15 (1965), 419-426.

85. CLARK, A. Homotopy commutativity and the Moore spectral sequence. Pacific J. Math. 15 (1965), 65-74.

86. CONLON, L. On the topology of EIII and EIV. Proc. Amer. Math. Soc. 16 (1965), 575-581.

87. CONLON, L. An application of the Bott suspension map to the topology of EIV. Pacific J. Math. 19 (1966), 411-428.

88. COPELAND, A.H., Jr. The Pontrjagin ring for certain loop spaces. Proc. Amer. Math. Soc. 7 (1956), 528-534.

89. COPELAND, A.H., Jr. On H-spaces with two non-trivial homotopy groups. Proc. Amer. Math. Soc. 8 (1957), 184-191.

90. COPELAND, A.H., Jr. Binary operations on sets of mapping classes. Michigan Math. J. 6 (1959), 7-23.

91. CURJEL, C.R. On the H-space structures of finite complexes. Comment. Math. Helv. 43 (1968), 1-17.

92. CURTIS, M. and MISLIN, G. Two new H-spaces. Bull. Amer. Math. Soc. 76 (1970), 851-852.

93. DOLD, A. and LASHOF, R. Principal quasi-fibrations and fibre homotopy equivalence of bundles. Illinois J. Math. 3 (1959), 285-305.

94. DOLD, A. and THOM, R. Quasifaserungen und unendliche symmetrische Produkte. Ann. of Math. 67 (1958), 239-281.

95. DOUGLAS, R.R. Homotopy-commutativity in H-spaces. Quart. J. Math. Oxford (2) 71 (1967), 263-283.

96. DOUGLAS, R.R., HILTON, P.J. and SIGRIST, F. H-spaces. Proc. Conf. Cat. and Hom. Alg. Seattle. Lecture Notes in Mathematics no. 92 (Springer Verlag, 1969).

97. DOUGLAS, R.R. and SIGRIST, F. Sphere-bundles over spheres which are H-spaces. Rendic. Acc. Naz. Lincei 44 (1968), 502-505.

98. DOUGLAS R.R. and SIGRIST, F. Homotopy-associative H-spaces which are sphere-bundles over spheres. Comment. Math. Helv. 44 (1969), 308-309.

99. DOUGLAS, R.R. and SIGRIST, F. Sphere bundles over spheres and H-spaces. Topology 8 (1969), 115-118.

100. DYER, E. and LASHOF, R. A topological proof of the Bott periodicity theorems. Ann. Mat. Pura Appl. (4) 54 (1961), 231-254.

101. DYER, E. and LASHOF, R. Homology of iterated loop spaces. Amer. J. Math. 84 (1962), 35-88.

102. DYNKIN, E.B. Homologies of compact Lie groups. Usp. Mat. Nauk. 8 (1953), 73-120. (Russian)

103. DYNKIN, E.B. Corrections to the paper, "Homologies of compact Lie groups". Usp. Mat. Nauk. 9 (1954), 233. (Russian)

104. ECKMANN, B. Über die Homotopiegruppen von Gruppenräumen. Comment. Math. Helv. 14 (1942), 234-256.

105. ECKMANN, B. and HILTON, P.J. A natural transformation in homotopy theory and a theorem of G.W. Whitehead. Math. Zeit. 82 (1963), 115-124.

106. FREUDENTHAL, H. Die Topologie der Lieschen Gruppen als algebraisches Phänomen I. Ann. of Math. 42 (1941), 1051-1074.

107. FUCHS, M. Verallgemeinerte Homotopie-Homomorphismen und klassifizierende Räume. Math. Ann. 161 (1965), 197-230.

108. GANEA, T., HILTON, P.J. and PETERSON, F.P. On the homotopy-commutativity of loop-spaces and suspensions. Topology 1 (1962), 133-141.

109. GITLER, S. Spaces fibred by H-spaces. Bol. Soc. Mat. Mexicana (2) 7 (1962), 71-84.

110. GITLER, S. On the Cartan suspension. Bol. Soc. Mat. Mexicana (2) 7 (1962), 85-91.

111. GITLER, S. The projective Stiefel manifolds II Applications. Topology 7 (1968), 47-53.

112. GITLER, S. The projective Stiefel manifolds II Applications (corrections). Topology 8 (1969), 93.

113. GITLER, S. and HANDEL, D. The projective Stiefel manifolds I. Topology 7 (1968), 39-46.

114. HALPERN, E. Twisted polynomial hyperalgebras. Mem. Amer. Math. Soc. 29 (1958).

115. HALPERN, E. On the structure of hyperalgebras. Class I Hopf algebras. Portugal Math. 17 (1958), 127-147.

116. HALPERN, E. The cohomology algebra of certain loop spaces. Proc. Amer. Math. Soc. 9 (1958), 808-817.

117. HALPERN, E. The cohomology of a space on which an H-space operates. Michigan Math. J. 6 (1959), 381-394.

118. HALPERN, E. On the primitivity of Hopf algebras over a field with prime characteristic. Proc. Amer. Math. Soc. 11 (1960), 117-126.

119. HARRIS, B. A generalization of H-spaces. Bull. Amer. Math. Soc. 66 (1960), 503-505.

120. HARRIS, B. On the homotopy groups of the classical groups. Ann. of Math. 74 (1961), 407-413.

121. HARRIS, B. Suspensions and characteristic maps for symmetric spaces. Ann. of Math. 76 (1962), 295-305.

122. HARRIS, B. Some calculations of homotopy groups of symmetric spaces. Trans. Amer. Math. Soc. 106 (1963), 174-184.

123. HARRIS, B. Torsion in Lie groups and related spaces. Topology 5 (1966), 347-354.

124. HARRIS, B. The K-theory of a class of homogeneous spaces. Trans. Amer. Math. Soc. 131 (1968), 323-332.

125. HELFENSTEIN, H.G. Analytic Hopf surfaces. Canad. J. Math. 14 (1962), 329-333.

126. HELFENSTEIN, H.G. Topological H-surfaces. Canad. J. Math. 17 (1965), 847-849.

127. HELFENSTEIN, H.G. Conformal types of H-surfaces. Proc. Third Brazilian Math. Colloq. (1967).

128. HILTON, P.J. On the homotopy groups of the union of spheres. J. London Math. Soc. 30 (1955), 154-172.

129. HILTON, P.J. Note on the Jacobi identity for Whitehead products. Proc. Cambridge Philos. Soc. 57 (1961), 180-182.

130. HILTON, P.J. Note on a theorem of Stasheff. Bull. Acad. Polon. Sci. 10 (1962), 127-130.

131. HILTON, P.J. Note on H-spaces and nilpotency. Bull. Acad. Polon. Sci. 11 (1963), 505-509.

132. HILTON, P.J. Nilpotency and H-spaces. Topology 3 (1964), suppl. 2, 161-176.

133. HILTON, P.J. Remark on loop spaces. Proc. Amer. Math. Soc. 15 (1964), 596-600.

134. HILTON, P.J. and ROITBERG, J. Note on principal S^3-bundles. Bull. Amer. Math. Soc. 74 (1968), 957-959.

135. HILTON, P.J. and ROITBERG, J. On principal S^3-bundles over spheres. Ann. of Math. 90 (1969), 91-107.

136. HODGKIN, L. On K-theory of Lie groups. Topology 6 (1967), 1-36.

137. HOFMANN, K.H. Topologische Loops. Math. Zeit. 70 (1958), 13-37.

138. HOFMANN, K.H. Topologische Doppelloops. Math. Zeit. 70 (1958), 213-230.

139. HOFMANN, K.H. Topologische Loops mit schwachen assoziativitaets-forderungen. Math. Zeit. 70 (1958), 125-155.

140. HOLZSAGER, R. H-spaces of category ≤ 2. Topology 9 (1970), 211-216.

141. HOO, C.S. A generalization of a theorem of Hilton. Canad. Math. Bull. 11 (1968), 663-669.

142. HOO, C.S. On the homotopy-commutativity of loop spaces and suspensions. Canad. J. Math. 20 (1968), 1531-1536.

143. HOO, C.S. Homotopical nilpotency of loop spaces. Canad. J. Math. 21 (1969), 479-484.

144. HOO, C.S. Nilpotency class of a map and Stasheff's criterion. Pacific J. Math. 28 (1969), 375-380.

145. HOO, C.S. Multiplication on spaces with comultiplication. Canad. Math. Bull. 12 (1969), 499-506.

146. HOO, C.S. and MAHOWALD, M.E. Some homotopy groups of Stiefel manifolds. Bull. Amer. Math. Soc. 71 (1965), 661-667.

147. HOPF, H. Über den Rang geschlossener Liescher Gruppen. Comment. Math. Helv. 13 (1940), 119-143.

148. HOPF, H. Über die Topologie der Gruppen-Mannigfaltigkeiten und ihre Verallgemeinerungen. Ann. of Math. 42 (1941), 22-52.

149. HÖSLI, H. Sur les H-espaces à deux générateurs cohomologiques. C.R. Acad. Sci. Paris 270 (1970), 746-749.

150. HU, Sze-tsen. Some homotopy properties of topological groups and homogeneous spaces, Ann. of Math. 49 (1948), 67-74.

151. HUBBUCK, J.R. Some results in the theory of H-spaces. Bull. Amer. Math. Soc. 74 (1968), 965-967.

152. HUBBUCK, J.R. A note on complex Stiefel manifolds. J. London Math. Soc. (2) 1 (1969), 85-89.

153. HUBBUCK, J.R. On homotopy-commutative H-spaces. Topology 8 (1969), 119-126.

154. HUBBUCK, J.R. Generalized cohomology operations and H-spaces of low rank. Trans. Amer. Math. Soc. 141 (1969), 335-360.

155. HUBBUCK, J.R. Automorphisms of polynomial algebras and homotopy-commutativity in H-spaces. Osaka J. Math. 6 (1969), 197-209.

156. HUBBUCK, J.R. Finitely generated cohomology Hopf algebras. Topology 9 (1970), 205-210.

157. HUSSEINI, S.Y. Constructions of the reduced product type. Topology 2 (1963), 213-237.

158. HUSSEINI, S.Y. On the representation of loop spaces as complexes of the reduced product type. Quart. J. Math. Oxford (2) 16 (1965), 13-31.

159. IWATA, K. Note on Postnikov invariants of a loop space. Tohoku Math. J. (2) 8 (1956), 329-332.

160. JAMES, I.M. Note on factor spaces. J. London Math. Soc. 28 (1953), 278-285.

161. JAMES, I.M. On the iterated suspension. Quart. J. Math. Oxford (2) 5 (1954), 1-10.

162. JAMES, I.M. Reduced product spaces. Ann. of Math. 62 (1955), 170-197.

163. JAMES, I.M. Symmetric functions of several variables, whose range and domain is a sphere. Bol. Soc. Mat. Mexicana (2) 1 (1956), 85-88.

164. JAMES, I.M. Commutative products on spheres. Proc. Cambridge Philos. Soc. 53 (1957), 63-68.

165. JAMES, I.M. Multiplication on spheres I. Proc. Amer. Math. Soc. 8 (1957), 192-196.

166. JAMES, I.M. Multiplication on spheres II. Trans. Amer. Math. Soc. 84 (1957), 545-558.

167. JAMES, I.M. On spaces with a multiplication. Pacific J. Math. 7 (1957), 1083-1100.

168. JAMES, I.M. Products on spheres. Mathematika 6 (1959), 1-13.

169. JAMES, I.M. On Lie groups and their homotopy groups. Proc. Cambridge Philos. Soc. 55 (1959), 244-247.

170. JAMES, I.M. On H-spaces and their homotopy groups. Quart. J. Math. Oxford (2) 11 (1960), 161-179.

171. JAMES, I.M. On sphere-bundles over spheres. Comment. Math. Helv. 35 (1961), 126-135.

172. JAMES, I.M. Quasigroups and topology. Math. Zeit. 84 (1964), 329-342.

173. JAMES, I.M. On homotopy-commutativity. Topology 6 (1967), 405-410.

174. JAMES, I.M. On the homotopy theory of the classical groups. An. Acad. Bras. Cienc. 39 (1967), 39-44.

175. JAMES, I.M. On fibre bundles and their homotopy groups. J. Math. Kyoto Univ. 9 (1969), 5-24.

176. JAMES, I.M. On the Bott suspension. J. Math. Kyoto Univ. 9 (1969), 161-188.

177. JAMES, I.M. and THOMAS, E. Homotopy-abelian topological groups. Topology 1 (1962), 237-240.

178. JAMES, I.M., THOMAS, E., TODA, H. and WHITEHEAD, G.W. On the symmetric square of a sphere. J. Math. Mech. 12 (1963), 771-776.

179. JAMES, I.M. and WHITEHEAD, J.H.C. Note on fibre spaces. Proc. London Math. Soc. (3) 4 (1954), 129-137.

180. KACHI, H. Homotopy groups of compact Lie groups E_6, E_7, E_8. Nagoya Math. J. 32 (1968), 109-140.

181. KAHN, D.W. A note on H-spaces and Postnikov systems of spheres. Proc. Amer. Math. Soc. 15 (1964), 300-307.

182. KERVAIRE, M.A. Some nonstable homotopy groups of Lie groups. Illinois J. Math. 4 (1960), 161-169.

183. KOJIMA, J. On the Pontrjagin product mod 2 of spinor groups. Mem. Fac. Sci. Kyusyu Univ. 11 (1957), 1-14.

184. KRAINES, D. Primitive chains and $H_*(\Omega X)$. Topology 8 (1969), 31-38.

185. KRISHNARAO, G.V. Unstable homotopy of O(n). Trans. Amer. Math. Soc. 127 (1967), 90-97.

186. KUIPER, N.H. The homotopy type of the unitary group of Hilbert space. Topology 3 (1965), 19-30.

187. KUMPEL, P.G., Jr. Lie groups and products of spheres. Proc. Amer. Math. Soc. 16 (1965), 1350-1356.

188. KUMPEL, P.G., Jr. On the homotopy groups of the exceptional Lie groups. Trans. Amer. Math. Soc. 120 (1965), 481-498.

189. LANDWEBER, P.S. On symmetric maps between spheres and equivariant K-theory. Topology 9 (1970), 55-61.

190. LEMAIRE, J-M. Un lemme sur l'homologie des certain espaces de lacets. C.R. Acad. Sci. Paris 269 (1969), 1122-1124.

191. LEMMENS, P.W.H. Homotopy theory of products of spheres, I, II. Proc. Royal Neth. Acad. Sci. A 72 (1969), 242-272.

192. LIU, Y-S. On ring-like spaces I. Advancement in Math. 3 (1957), 404-408. (Chinese. English summary)

193. LOIBEL, G.F. Multiplications on products of spheres. An. Acad. Bras. Cienc. 31 (1959), 161-162.

194. LOIBEL, G.F. Über Multiplikationen im n-dimensionalen Torus. Ann. Mat. Pura Appl. (4) 52 (1960), 27-33.

195. LOIBEL, G.F. On topological quasi-groups. Bol. Soc. Mat. São Paulo 13 (1958), 1-42. (Portuguese)

196. LUNDELL, A.T. The embeddings $O(n) \subset Sp(n)$ and $U(n) \subset Sp(n)$ and a Samelson product. Michigan Math. J. 13 (1966), 133-145.

197. LUNDELL, A.T. Homotopy periodicity of the classical Lie groups. Proc. Amer. Math. Soc. 18 (1967), 683-690.

198. LUNDELL, A.T. A Bott map for non-stable homotopy of the unitary group. Topology 8 (1969), 209-217.

199. McCARTY, G.S., Jr. Products between homotopy groups and the J-morphism. Quart. J. Math. Oxford (2) 15 (1964), 362-370.

200. McCARTY, G.S., Jr. The value of J at a Samelson product. Proc. Amer. Math. Soc. 19 (1968), 164-167.

201. McCORD, M.C. Classifying spaces and infinite symmetric products. Trans. Amer. Math. Soc. 146 (1969), 273-298.

202. MAHOWALD, M.E. On the order of the image of J. Topology 6 (1967), 371-378.

203. MAHOWALD, M.E. On the metastable homotopy of O(n). Proc. Amer. Math. Soc. 19 (1968), 639-641.

204. MAHOWALD, M.E. A Samelson product in SO(2n). Bol. Soc. Mat. Mexicana (2) 10 (1965), 80-83.

205. MALCEV, A.I. Analytic loops. Mat. Sb. 36 (77) (1954), 569-576.

206. MASSEY, W.S. and UEHARA, H. The Jacobi identity for Whitehead products. Algebraic geometry and topology. A symposium in honour of S. Lefschetz, 361-377. (Princeton University Press, 1957).

207. MATSUNAGA, H. On the homotopy groups of Stiefel manifolds. Mem. Fac. Sci. Kyusyu Univ. 13 (1959), 152-156.

208. MATSUNAGA, H. The homotopy groups $\pi_{2n+i}(U(n))$ for i = 3, 4 and 5. Mem. Fac. Sci. Kyusyu Univ. 15 (1961/62), 72-81.

209. MATSUNAGA, H. On the groups $\pi_{2n+7}(U(n))$, odd primary components. Mem. Fac. Sci. Kyusyu Univ. 16 (1962), 66-74.

210. MATSUNAGA, H. Correction to the preceding paper and note on the James number. Mem. Fac. Sci. Kyusyu Univ. 16 (1962), 60-61.

211. MATSUNAGA, H. Applications of functional cohomology operations to the calculus of $\pi_{2n+i}(U(n))$ for $i = 6$ and 7, $n \geqq 4$. Mem. Fac. Sci. Kyusyu Univ. 17 (1963), 29-62.

212. MATSUNAGA, H. Unstable homotopy groups of unitary groups (odd primary components). Osaka J. Math. 1 (1964), 15-24.

213. MAY, J.P. Categories of spectra and infinite loop spaces. Category theory, homology theory and their applications III. Lecture Notes in Mathematics no. 99 (Springer Verlag, 1969), 448-479.

214. MILGRAM, J.R. Iterated loop spaces. Ann. of Math. 84 (1966), 386-403.

215. MILGRAM, J.R. The bar construction and abelian H-spaces. Illinois J. Math. 11 (1967), 242-250.

216. MILLER, C.E. The topology of rotation groups. Ann. of Math. 57 (1953), 90-114.

217. MILNOR, J. Construction of universal bundles I. Ann. of Math. 63 (1956), 272-284.

218. MILNOR, J. Construction of universal bundles II. Ann. of Math. 63 (1956), 430-436.

219. MILNOR, J. On the Whitehead homomorphism J. Bull. Amer. Math. Soc. 64 (1958), 79-82.

220. MILNOR, J. and MOORE, J.C. On the structure of Hopf algebras. Ann. of Math. 81 (1965), 211-264.

221. MIMURA, M. Quelques groupes d'homotopie metastables des espaces symmetriques Sp(n) et U(2n)/Sp(n). C.R. Acad. Sci. Paris 262 (1966), 20-21.

222. MIMURA, M. The homotopy groups of Lie groups of low rank. J. Math. Kyoto Univ. 6 (1967), 131-176.

223. MIMURA, M. The number of multiplications on SU(3) and Sp(2). Trans. Amer. Math. Soc. 146 (1969), 473-492.

224. MIMURA, M. and TODA, H. Homotopy groups of SU(3), SU(4) and Sp(2). J. Math. Kyoto Univ. 3 (1963/64), 217-250.

225. MIMURA, M. and TODA, H. Homotopy groups of symplectic groups. J. Math. Kyoto Univ. 3 (1964), 251-273.

226. MOORE, J.C. and SMITH, L. Hopf algebras and multiplicative fibrations I. Amer. J. Math. 90 (1968), 752-780.

227. MOORE, J.C. and SMITH, L. Hopf algebras and multiplicative fibrations II. Amer. J. Math. 90 (1968), 1113-1150.

228. NAKAOKA, M. and TODA, H. On Jacobi identity for Whitehead products. J. Inst. Polytech. Osaka City Univ. 5 (1954), 1-13.

229. NAYLOR, C.M. Multiplications on SO(3). Michigan Math. J. 13 (1966), 27-31.

230. NISHIDA, G. Cohomology operations in iterated loop spaces. Proc. Japan Acad. 44 (1968), 104-109.

231. NORMAN, C.W. Homotopy loops. Topology 2 (1963), 23-43.

232. NOVIKOV, S.P. Homotopy properties of the group of diffeomorphisms of the sphere. Dokl. Akad. Nauk. SSSR 148 (1963), 32-35. (Russian)

233. OCHIAI, S. On the type of an associative H-space. Proc. Japan Acad. 45 (1969), 92-94.

234. OGUCHI, K. 2-primary components of the homotopy groups of some Lie groups. Proc. Japan Acad. 38 (1962), 235-238.

235. OGUCHI, K. 2-primary components of the homotopy groups of spheres and Lie groups. Proc. Japan Acad. 38 (1962), 619-620.

236. OGUCHI, K. Generators of 2-primary components of homotopy groups of spheres, unitary groups and symplectic groups. J. Fac. Sci. Univ. Tokyo 11 (1964), 65-111.

237. OGUCHI, K. Homotopy groups of Sp(n)/Sp(n-2). J. Fac. Sci. Univ. Tokyo 16 (1969), 179-201.

238. O'NEILL, R.C. On H-spaces that are CW-complexes I. Illinois J. Math. 8 (1964), 280-290.

239. O'NEILL, R.C. Retracts and retractile subcomplexes. Topology 5 (1966), 191-202.

240. ONISCIK, A.L. On cohomologies of spaces of paths. Mat. Sb. 44 (86) (1958), 3-52. (Russian)

241. ONISCIK, A.L. Torsion of the special Lie groups. Mat. Sb. 51 (93) (1960), 273-276.

242. ONO, Y. Notes on the Lusternik-Schirelmann category and H-spaces. Bull. Kyoto Univ. Education no. 34 (1969), 1-4.

243. PAECHTER, G.F. The groups $\pi_r(V_{n,m})$ I. Quart. J. Math. Oxford (2) 7 (1956), 249-268.

244. PAECHTER, G.F. The groups $\pi_r(V_{n,m})$ II. Quart. J. Math. Oxford (2) 9 (1958), 8-27.

245. PAECHTER, G.F. The groups $\pi_r(V_{n,m})$ III. Quart. J. Math. Oxford (2) 10 (1959), 17-37.

246. PAECHTER, G.F. The groups $\pi_r(V_{n,m})$ IV. Quart. J. Math. Oxford (2) 10 (1959), 241-260.

247. PAECHTER, G.F. The groups $\pi_r(V_{n,m})$ V. Quart. J. Math. Oxford (2) 11 (1960), 1-16.

248. PETERSON, F.P. A note on H-spaces. Bol. Soc. Mat. Mexicana (2) 4 (1959), 30-31.

249. PETRIE, T. The K-theory of the projective unitary groups. Bull. Amer. Math. Soc. 72 (1966), 875-878.

250. PETRIE, T. The K-theory of the projective unitary groups. Topology 6 (1967), 103-116.

251. PETRIE, T. The weakly-complex bordism of Lie groups. Ann. of Math. 88 (1968), 371-402.

252. PONTRJAGIN, L. Homologies in compact Lie groups. Mat. Sb. 6 (48) (1939), 389-422. (English. Russian summary)

253. PONTRJAGIN, L. Über die topologische Struktur der Lieschen Gruppen. Comment. Math. Helv. 13 (1941), 277-283.

254. PORTER, G.J. Homotopical nilpotence of S^3. Proc. Amer. Math. Soc. 15 (1964), 681-682.

255. PORTER, G.J. On the homotopy groups of the wedge of spheres. Amer. J. Math. 87 (1965), 297-314.

256. PUTNAM, R. and WINTNER, A. The connectedness of the orthogonal group in Hilbert space. Proc. Nat. Sci. USA 37 (1951), 110-112.

257. PUTNAM, R. and WINTNER, A. The orthogonal group in Hilbert space. Amer. J. Math. 74 (1952), 52-78.

258. RAMANUJAM, S. Topology of classical groups, Osaka J. Math. 6 (1969), 243-249.

259. RAMANUJAM, S. On Stiefel manifolds, J. Math. Soc. Japan 21 (1969), 543-548.

260. REES, E. Multiplications on projective spaces. Michigan Math. J. 16 (1969), 297-302.

261. ROTHENBERG, M. The J functor and the nonstable homotopy groups of the unitary groups. Proc. Amer. Math. Soc. 15 (1964), 264-271.

262. ROTHENBERG, M. and STEENROD, N.E. The cohomology of classifying spaces of H-spaces. Bull. Amer. Math. Soc. 71 (1965), 872-875.

263. SAITO, Y. On the homotopy groups of Stiefel manifolds. J. Inst. Poly. Osaka City Univ. 6 (1955), 39-45.

264. SAITO, Y., TODA, H. and YOKOTA, I. Note on the generator of $\pi_7(SO(n))$. Mem. Coll. Sci. Univ. Kyoto 30 (1957), 227-230.

265. SAMELSON, H. Über die Sphären, die als Gruppenräume auftreten. Comment. Math. Helv. 13 (1940), 144-155.

266. SAMELSON, H. Beiträge zur Topologie der Gruppen-Mannigfaltigkeiten. Ann. of Math. 42 (1941), 1091-1137.

267. SAMELSON, H. Topology of Lie groups. Bull. Amer. Math. Soc. 58 (1952), 2-37.

268. SAMELSON, H. On the relation between the Whitehead and the Pontrjagin product. Amer. J. Math. 75 (1953), 744-752.

269. SAMELSON, H. Groups and spaces of loops. Comment. Math. Helv. 28 (1954), 278-287.

270. SCHEERER, H. Homotopieäquivalente kompakte Liesche Gruppen. Topology 7 (1968), 227-232.

271. SEGAL, G.B. Classifying spaces and spectral sequences. Inst. Hautes Etudes Sci. Publ. Math. no. 34 (1968), 105-112.

272. SEGAL, G.B. Equivariant contractibility of the general linear group of Hilbert space. Bull. London Math. Soc. 1 (1969), 329-331.

273. SERRE, J-P. Groupes d'homotopie et classes de groupes abéliens. Ann. of Math. 58 (1953), 258-294.

274. SIBSON, R. Existence theorem for H-space inverses. Proc. Cambridge Philos. Soc. 65 (1969), 19-21.

275. SIEGEL, J. G-spaces, H-spaces and W-spaces. Pacific J. Math. 31 (1969), 209-214.

276. SIGRIST, F. Groupes d'homotopie des variétés de Stiefel complexes. Comment. Math. Helv. 43 (1968), 121-131.

277. SIGRIST, F. Determination des groupes d'homotopie $\pi_{2k+7}(U_{k+m,m})$. C.R. Acad. Sci. Paris 269 (1969), 1061-1062.

278. SINGER, W.M. Connective fiberings over BU and U. Topology 7 (1968), 271-303.

279. SLIFKER, J.F. Exotic multiplications on S^3. Quart. J. Math. Oxford (2) 16 (1965), 322-359.

280. SMITH, L. Cohomology of $\Omega(G/U)$. Proc. Amer. Math. Soc. 19 (1968), 399-404.

281. SMITH, L. On the type of an associative H-space of rank two. Tohoku Math. J. 20 (1968), 511-515.

282. SMITH, L. On the type of an associative H-space of rank three. Proc. Japan Acad. 44 (1968), 811-815.

283. SMITH, L. On the relation between spherical and primitive homology classes in topological groups. Topology 8 (1969), 69-80.

284. SMITH, L. Lectures on the Eilenberg-Moore spectral sequence. Lecture Notes in Mathematics no. 134 (Springer Verlag, 1970).

285. SMITH, L. and STONG, R.E. The structure of BSC. Inventiones Math. 5 (1968), 138-159.

286. SPANIER, E.H. and WHITEHEAD, J.H.C. On fibre spaces in which the fibre is contractible. Comment. Math. Helv. 29 (1955), 1-8.

287. STASHEFF, J. On the space-of-loops isomorphism. Proc. Amer. Math. Soc. 10 (1959), 987-993.

288. STASHEFF, J. On homotopy abelian H-spaces. Proc. Cambridge Philos. Soc. 57 (1961), 734-745.

289. STASHEFF, J. On extensions of H-spaces. Trans. Amer. Math. Soc. 105 (1962), 126-135.

290. STASHEFF, J. Homotopy associativity of H-spaces I. Trans. Amer. Math. Soc. 108 (1963), 275-292.

291. STASHEFF, J. Homotopy associativity of H-spaces II. Trans. Amer. Math. Soc. 108 (1963), 293-312.

292. STASHEFF, J. Manifolds of the homotopy type of (non-Lie) groups. Bull. Amer. Math. Soc. 75 (1969), 998-1000.

293. STASHEFF, J. Torsion in BBSO. Pacific J. Math. 28 (1969), 677-680.

294. STASHEFF, J. H-spaces from a homotopy point of view. Lecture Notes in Mathematics no. 161 (Springer Verlag, 1970).

295. STEENROD, N.E. Cohomology operations. Lectures by N.E. Steenrod written and revised by D.B.A. Epstein. Annals of Mathematics Studies no. 50. (Princeton University Press, 1962).

296. STEENROD, N.E. Milgram's classifying space of a topological group. Topology 7 (1968), 349-368.

297. STEER, B. Extensions of mappings into H-spaces. Proc. London Math. Soc. (3) 13 (1963), 219-272.

298. STEER, B. Generalized Whitehead products. Quart. J. Math. Oxford (2) 14 (1963), 29-40.

299. STEER, B. Transgression in sphere-bundles. Topology 3 (1963), 1-10.

300. STEER, B. Note on the Hurewicz homomorphism in the space of homotopy equivalences of S^n preserving basepoint. Quart. J. Math. Oxford (2) 19 (1968), 245-248.

301. STEIN, S.K. On the foundations of quasigroups. Trans. Amer. Math. Soc. 85 (1957), 228-256.

302. SUGAWARA, M. On the homotopy groups of rotation groups. Math. J. Okayama Univ. 3 (1953), 11-21.

303. SUGAWARA, M. Some remarks on homotopy groups of rotation groups. Math. J. Okayama Univ. 3 (1954), 129-133.

304. SUGAWARA, M. H-spaces and spaces of loops. Math. J. Okayama Univ. 5 (1955), 5-11.

305. SUGAWARA, M. On fibres of fibre spaces whose total space is contractible. Math. J. Okayama Univ. 5 (1956), 127-131.

306. SUGAWARA, M. On a condition that a space is an H-space. Math. J. Okayama Univ. 6 (1957), 109-129.

307. SUGAWARA, M. A condition that a space is group-like. Math. J. Okayama Univ. 7 (1957), 123-149.

308. SUGAWARA, M. Some remarks on homotopy equivalences and H-spaces. Math. J. Okayama Univ. 8 (1958), 125-131.

309. SUGAWARA, M. Note on H-spaces. Proc. Japan Acad. 36 (1960), 598-600.

310. SUGAWARA, M. On the homotopy-commutativity of groups and loop spaces. Mem. Coll. Sci. Univ. Kyoto 33 (1960/61), 257-269.

311. SUZUKI, H. On the Eilenberg-MacLane invariants of loop spaces. J. Math. Soc. Japan 8 (1956), 93-101.

312. SUZUKI, H. Multiplications in Postnikov systems and their applications. Tohuku Math. J. (2) 12 (1960), 389-399.

313. SUZUKI, H. Remarks on the Multiplications in Postnikov systems. Mem. Fac. Sci. Kyusyu Univ. 17 (1963), 200-201.

314. SUZUKI, H. Hopf space mod $\mathcal{C}$. Mem. Fac. Sci. Kyusyu Univ. 17 (1963), 1-9.

315. SVARC, A.S. Homologies of the spinor group. Dokl. Akad. Nauk. SSSR 104 (1955), 26-29. (Russian)

316. TAKEUCHI, M. On Pontrjagin classes of compact symmetric spaces. J. Fac. Sci. Univ. Tokyo 9 (1962), 313-328.

317. TAKEUCHI, M. Cell decompositions and Morse inequalities on certain symmetric spaces. J. Fac. Sci. Univ. Tokyo 12 (1965), 81-102.

318. THOMAS, E. On the cohomology groups of the classifying space for the stable spinor group. Bol. Soc. Mat. Mexicana (2) 7 (1962), 57-69.

319. THOMAS, E. On the mod 2 cohomology of certain H-spaces. Comment. Math. Helv. 37 (1962/63), 132-140.

320. THOMAS, E. Steenrod squares and H-spaces. Ann. of Math. 77 (1963), 306-317.

321. THOMAS, E. Exceptional Lie groups and Steenrod squares. Michigan Math. J. 11 (1964), 151-156.

322. THOMAS, E. Steenrod squares and H-spaces II. Ann. of Math. 81 (1965), 473-495.

323. TODA, H. Quelques tables des groupes d'homotopie des groupes de Lie. C.R. Acad. Sci. Paris 241 (1955), 922-923.

324. TODA, H. Non-existence of mappings of S^{31} into S^{16} with Hopf invariant 1. J. Inst. Poly. Osaka City Univ. 8 (1957), 31-34.

325. TODA, H. A topological proof of theorems of Bott and Borel-Hirzebruch for homotopy groups of unitary groups. Mem. Coll. Sci. Univ. Kyoto 32 (1959), 103-119.

326. TODA, H. On unstable homotopy of spheres and classical groups. Proc. Nat. Acad. Sci. USA 46 (1960), 1102-1105.

327. TODA, H. Vector fields on spheres. Bull. Amer. Math. Soc. 67 (1961), 408-412.

328. TODA, H. Composition methods in homotopy groups of spheres. Annals of Mathematics Studies no. 49. (Princeton University Press, 1962).

329. TODA, H. On homotopy groups of S^3-bundles over spheres. J. Math. Kyoto Univ. 2 (1963), 193-207.

330. UCCI, J.J. On symmetric maps of spheres. Inventiones Math. 5 (1968), 8-18.

331. WADA, H. Note on some mapping spaces. Tohuku Math. J. (2) 10 (1958), 143-145.

332. WEST, R.W. Weak H-spaces. J. Math. Mech. 17 (1967), 421-431.

333. WHITEHEAD, G.W. On the homotopy groups of spheres and rotation groups. Ann. of Math. 43 (1942), 634-640.

334. WHITEHEAD, G.W. Homotopy properties of the real orthogonal groups. Ann. of Math. 43 (1942), 132-146.

335. WHITEHEAD, G.W. On mappings into group-like spaces. Comment. Math. Helv. 28 (1954), 320-328.

336. WHITEHEAD, G.W. Note on a theorem of Sugawara. Bol. Soc. Mat. Mexicana (2) 4 (1959), 33-41.

337. WILLIAMS, F.D. A characterization of spaces with vanishing Whitehead products. Bull. Amer. Math. Soc. 74 (1968), 497-499.

338. WILLIAMS, F.D. Higher homotopy-commutativity. Trans. Amer. Math. Soc. 139 (1969), 191-206.

339. WOOD, R. Banach algebras and Bott periodicity. Topology 4 (1965/66), 371-389.

340. YEN, Chih-ta. Sur les polynomes de Poincaré des groupes simples exceptionels. C.R. Acad. Sci. Paris 228 (1949), 628-630.

341. YOKOTA, I. On the cell structures of SU(n) and Sp(n). Proc. Japan Acad. 31 (1955), 673-677.

342. YOKOTA, I. On the cells of symplectic groups. Proc. Japan Acad. 32 (1956), 399-400.

343. YOKOTA, I. On the cellular decomposition of unitary groups. J. Inst. Poly. Osaka City Univ. 7 (1956), 39-49.

344. YOKOTA, I. On the homology of classical Lie groups. J. Inst. Poly. Osaka City Univ. 8 (1957), 93-120.

345. YOKOTA, I. On some homogeneous spaces of classical Lie groups. J. Inst. Poly. Osaka City Univ. 9 (1958), 29-35.

346. ZABRODSKY, A. Homotopy associativity and finite CW-complexes. Topology 9 (1970), 121-128.

347. ZEEMAN, E.C. A note on a theorem of Armand Borel. Proc. Cambridge Philos. Soc. 54 (1958), 396-398.

Lecture Notes in Mathematics

Vol. 72: The Syntax and Semantics of Infinitary Languages. Edited by J. Barwise. IV, 268 pages. 1968. DM 18,– / $ 5.00

Vol. 73: P. E. Conner, Lectures on the Action of a Finite Group. IV, 123 pages. 1968. DM 10,– / $ 2.80

Vol. 74: A. Fröhlich, Formal Groups. IV, 140 pages. 1968. DM 12,– / $ 3.30

Vol. 75: G. Lumer, Algèbres de fonctions et espaces de Hardy. VI, 80 pages. 1968. DM 8,– / $ 2.20

Vol. 76: R. G. Swan, Algebraic K-Theory. IV, 262 pages. 1968. DM 18,– / $ 5.00

Vol. 77: P.-A. Meyer, Processus de Markov: la frontière de Martin. IV, 123 pages. 1968. DM 10,– / $ 2.80

Vol. 78: H. Herrlich, Topologische Reflexionen und Coreflexionen. XVI, 166 Seiten. 1968. DM 12,– / $ 3.30

Vol. 79: A. Grothendieck, Catégories Cofibrées Additives et Complexe Cotangent Relatif. IV, 167 pages. 1968. DM 12,– / $ 3.30

Vol. 80: Seminar on Triples and Categorical Homology Theory. Edited by B. Eckmann. IV, 398 pages. 1969. DM 20,– / $ 5.50

Vol. 81: J.-P. Eckmann et M. Guenin, Méthodes Algébriques en Mecanique Statistique. VI, 131 pages. 1969. DM 12,– / $ 3.30

Vol. 82: J. Wloka, Grundräume und verallgemeinerte Funktionen. VIII, 131 Seiten. 1969. DM 12,– / $ 3.30

Vol. 83: O. Zariski, An Introduction to the Theory of Algebraic Surfaces. IV, 100 pages. 1969. DM 8,– / $ 2.20

Vol. 84: H. Lüneburg, Transitive Erweiterungen endlicher Permutationsgruppen. IV, 119 Seiten. 1969. DM 10,– / $ 2.80

Vol. 85: P. Cartier et D. Foata, Problèmes combinatoires de commutation et réarrangements. IV, 88 pages. 1969. DM 8,– / $ 2.20

Vol. 86: Category Theory, Homology Theory and their Applications I. Edited by P. Hilton. VI, 216 pages. 1969. DM 16,– / $ 4.40

Vol. 87: M. Tierney, Categorical Constructions in Stable Homotopy Theory. IV, 65 pages. 1969. DM 6,– / $ 1.70

Vol. 88: Séminaire de Probabilités III. IV, 229 pages. 1969. DM 18,– / $ 5.00

Vol. 89: Probability and Information Theory. Edited by M. Behara, K. Krickeberg and J. Wolfowitz. IV, 256 pages. 1969. DM 18,– / $ 5.00

Vol. 90: N. P. Bhatia and O. Hajek, Local Semi-Dynamical Systems. II, 157 pages. 1969. DM 14,– / $ 3.90

Vol. 91: N. N. Janenko, Die Zwischenschrittmethode zur Lösung mehrdimensionaler Probleme der mathematischen Physik. VIII, 194 Seiten. 1969. DM 16,80 / $ 4.70

Vol. 92: Category Theory, Homology Theory and their Applications II. Edited by P. Hilton. V, 308 pages. 1969. DM 20,– / $ 5.50

Vol. 93: K. R. Parthasarathy, Multipliers on Locally Compact Groups. III, 54 pages. 1969. DM 5,60 / $ 1.60

Vol. 94: M. Machover and J. Hirschfeld, Lectures on Non-Standard Analysis. VI, 79 pages. 1969. DM 6,– / $ 1.70

Vol. 95: A. S. Troelstra, Principles of Intuitionism. II, 111 pages. 1969. DM 10,– / $ 2.80

Vol. 96: H.-B. Brinkmann und D. Puppe, Abelsche und exakte Kategorien, Korrespondenzen. V, 141 Seiten. 1969. DM 10,– / $ 2.80

Vol. 97: S. O. Chase and M. E. Sweedler, Hopf Algebras and Galois theory. II, 133 pages. 1969. DM 10,– / $ 2.80

Vol. 98: M. Heins, Hardy Classes on Riemann Surfaces. III, 106 pages. 1969. DM 10,– / $ 2.80

Vol. 99: Category Theory, Homology Theory and their Applications III. Edited by P. Hilton. IV, 489 pages. 1969. DM 24,– / $ 6.60

Vol. 100: M. Artin and B. Mazur, Etale Homotopy. II, 196 Seiten. 1969. DM 12,– / $ 3.30

Vol. 101: G. P. Szegö et G. Treccani, Semigruppi di Trasformazioni Multivoche. VI, 177 pages. 1969. DM 14,– / $ 3.90

Vol. 102: F. Stummel, Rand- und Eigenwertaufgaben in Sobolewschen Räumen. VIII, 386 Seiten. 1969. DM 20,– / $ 5.50

Vol. 103: Lectures in Modern Analysis and Applications I. Edited by C. T. Taam. VII, 162 pages. 1969. DM 12,– / $ 3.30

Vol. 104: G. H. Pimbley, Jr., Eigenfunction Branches of Nonlinear Operators and their Bifurcations. II, 128 pages. 1969. DM 10,– / $ 2.80

Vol. 105: R. Larsen, The Multiplier Problem. VII, 284 pages. 1969. DM 18,– / $ 5.00

Vol. 106: Reports of the Midwest Category Seminar III. Edited by S. Mac Lane. III, 247 pages. 1969. DM 16,– / $ 4.40

Vol. 107: A. Peyerimhoff, Lectures on Summability. III, 111 pages. 1969. DM 8,– / $ 2.20

Vol. 108: Algebraic K-Theory and its Geometric Applications. Edited by R. M. F. Moss and C. B. Thomas. IV, 86 pages. 1969. DM 6,– / $ 1.70

Vol. 109: Conference on the Numerical Solution of Differential Equations. Edited by J. Ll. Morris. VI, 275 pages. 1969. DM 18,– / $ 5.00

Vol. 110: The Many Facets of Graph Theory. Edited by G. Chartrand and S. F. Kapoor. VIII, 290 pages. 1969. DM 18,– / $ 5.00

Vol. 111: K. H. Mayer, Relationen zwischen charakteristischen Zahlen. III, 99 Seiten. 1969. DM 8,– / $ 2.20

Vol. 112: Colloquium on Methods of Optimization. Edited by N. N. Moiseev. IV, 293 pages. 1970. DM 18,– / $ 5.00

Vol. 113: R. Wille, Kongruenzklassengeometrien. III, 99 Seiten. 1970. DM 8,– / $ 2.20

Vol. 114: H. Jacquet and R. P. Langlands, Automorphic Forms on GL (2). VII, 548 pages. 1970. DM 24,– / $ 6.60

Vol. 115: K. H. Roggenkamp and V. Huber-Dyson, Lattices over Orders I. XIX, 290 pages. 1970. DM 18,– / $ 5.00

Vol. 116: Séminaire Pierre Lelong (Analyse) Année 1969. IV, 195 pages. 1970. DM 14,– / $ 3.90

Vol. 117: Y. Meyer, Nombres de Pisot, Nombres de Salem et Analyse Harmonique. 63 pages. 1970. DM 6,– / $ 1.70

Vol. 118: Proceedings of the 15th Scandinavian Congress, Oslo 1968. Edited by K. E. Aubert and W. Ljunggren. IV, 162 pages. 1970. DM 12,– / $ 3.30

Vol. 119: M. Raynaud, Faisceaux amples sur les schémas en groupes et les espaces homogènes. III, 219 pages. 1970. DM 14,– / $ 3.90

Vol. 120: D. Siefkes, Buchi's Monadic Second Order Successor Arithmetic. XII, 130 Seiten. 1970. DM 12,– / $ 3.30

Vol. 121: H. S. Bear, Lectures on Gleason Parts. III, 47 pages. 1970. DM 6,– / $ 1.70

Vol. 122: H. Zieschang, E. Vogt und H.-D. Coldewey, Flächen und ebene diskontinuierliche Gruppen. VIII, 203 Seiten. 1970. DM 16,– / $ 4.40

Vol. 123: A. V. Jategaonkar, Left Principal Ideal Rings. VI, 145 pages. 1970. DM 12,– / $ 3.30

Vol. 124: Séminare de Probabilités IV. Edited by P. A. Meyer. IV, 282 pages. 1970. DM 20,– / $ 5.50

Vol. 125: Symposium on Automatic Demonstration. V, 310 pages. 1970. DM 20,– / $ 5.50

Vol. 126: P. Schapira, Théorie des Hyperfonctions. XI, 157 pages. 1970. DM 14,– / $ 3.90

Vol. 127: I. Stewart, Lie Algebras. IV, 97 pages. 1970. DM 10,– / $ 2.80

Vol. 128: M. Takesaki, Tomita's Theory of Modular Hilbert Algebras and its Applications. II, 123 pages. 1970. DM 10,– / $ 2.80

Vol. 129: K. H. Hofmann, The Duality of Compact Semigroups and C*-Bigebras. XII, 142 pages. 1970. DM 14,– / $ 3.90

Vol. 130: F. Lorenz, Quadratische Formen über Körpern. II, 77 Seiten. 1970. DM 8,– / $ 2.20

Vol. 131: A Borel et al., Seminar on Algebraic Groups and Related Finite Groups. VII, 321 pages. 1970. DM 22,– / $ 6.10

Vol. 132: Symposium on Optimization. III, 348 pages. 1970. DM 22,– / $ 6.10

Vol. 133: F. Topsøe, Topology and Measure. XIV, 79 pages. 1970. DM 8,– / $ 2.20

Vol. 134: L. Smith, Lectures on the Eilenberg-Moore Spectral Sequence. VII, 142 pages. 1970. DM 14,– / $ 3.90

Vol. 135: W. Stoll, Value Distribution of Holomorphic Maps into Compact Complex Manifolds. II, 267 pages. 1970. DM 18,– / $ 5.00

Vol. 136: M. Karoubi et al., Séminaire Heidelberg-Saarbrücken-Strasbourg sur la K-Théorie. IV, 264 pages. 1970. DM 18,– / $ 5.00

Vol. 137: Reports of the Midwest Category Seminar IV. Edited by S. MacLane. III, 139 pages. 1970. DM 12,– / $ 3.30

Vol. 138: D. Foata et M. Schützenberger, Théorie Géométrique des Polynômes Eulériens. V, 94 pages. 1970. DM 10,– / $ 2.80

Vol. 139: A. Badrikian, Séminaire sur les Fonctions Aléatoires Linéaires et les Mesures Cylindriques. VII, 221 pages. 1970. DM 18,– / $ 5.00

Vol. 140: Lectures in Modern Analysis and Applications II. Edited by C. T. Taam. VI, 119 pages. 1970. DM 10,– / $ 2.80

Vol. 141: G. Jameson, Ordered Linear Spaces. XV, 194 pages. 1970. DM 16,– / $ 4.40

Vol. 142: K. W. Roggenkamp, Lattices over Orders II. V, 388 pages. 1970. DM 22,– / $ 6.10

Vol. 143: K. W. Gruenberg, Cohomological Topics in Group Theory. XIV, 275 pages. 1970. DM 20,– / $ 5.50

Lecture Notes in Mathematics – Lecture Notes in Physics

Vol. 144: Seminar on Differential Equations and Dynamical Systems, II. Edited by J. A. Yorke. VIII, 268 pages. 1970. DM 20,– / $ 5.50

Vol. 145: E. J. Dubuc, Kan Extensions in Enriched Category Theory. XVI, 173 pages. 1970. DM 16,– / $ 4.40

Vol. 146: A. B. Altman and S. Kleiman, Introduction to Grothendieck Duality Theory. II, 192 pages. 1970. DM 18,– / $ 5.00

Vol. 147: D. E. Dobbs, Cech Cohomological Dimensions for Commutative Rings. VI, 176 pages. 1970. DM 16,– / $ 4.40

Vol. 148: R. Azencott, Espaces de Poisson des Groupes Localement Compacts. IX, 141 pages. 1970. DM 14,– / $ 3.90

Vol. 149: R. G. Swan and E. G. Evans, K-Theory of Finite Groups and Orders. IV, 237 pages. 1970. DM 20,– / $ 5.50

Vol. 150: Heyer, Dualität lokalkompakter Gruppen. XIII, 372 Seiten. 1970. DM 20,– / $ 5.50

Vol. 151: M. Demazure et A. Grothendieck, Schémas en Groupes I. (SGA 3). XV, 562 pages. 1970. DM 24,– / $ 6.60

Vol. 152: M. Demazure et A. Grothendieck, Schémas en Groupes II. (SGA 3). IX, 654 pages. 1970. DM 24,– / $ 6.60

Vol. 153: M. Demazure et A. Grothendieck, Schémas en Groupes III. (SGA 3). VIII, 529 pages. 1970. DM 24,– / $ 6.60

Vol. 154: A. Lascoux et M. Berger, Variétés Kähleriennes Compactes. VII, 83 pages. 1970. DM 8,– / $ 2.20

Vol. 155: Several Complex Variables I, Maryland 1970. Edited by J. Horváth. IV, 214 pages. 1970. DM 18,– / $ 5.00

Vol. 156: R. Hartshorne, Ample Subvarieties of Algebraic Varieties. XIV, 256 pages. 1970. DM 20,– / $ 5.50

Vol. 157: T. tom Dieck, K. H. Kamps und D. Puppe, Homotopietheorie. VI, 265 Seiten. 1970. DM 20,– / $ 5.50

Vol. 158: T. G. Ostrom, Finite Translation Planes. IV. 112 pages. 1970. DM 10,– / $ 2.80

Vol. 159: R. Ansorge und R. Hass. Konvergenz von Differenzenverfahren für lineare und nichtlineare Anfangswertaufgaben. VIII, 145 Seiten. 1970. DM 14,– / $ 3.90

Vol. 160: L. Sucheston, Constributions to Ergodic Theory and Probability. VII, 277 pages. 1970. DM 20,– / $ 5.50

Vol. 161: J. Stasheff, H-Spaces from a Homotopy Point of View. VI, 95 pages. 1970. DM 10,– / $ 2.80

Vol. 162: Harish-Chandra and van Dijk, Harmonic Analysis on Reductive p-adic Groups. IV, 125 pages. 1970. DM 12,– / $ 3.30

Vol. 163: P. Deligne, Equations Différentielles à Points Singuliers Réguliers. III, 133 pages. 1970. DM 12,– / $ 3.30

Vol. 164: J. P. Ferrier, Seminaire sur les Algebres Completes. II, 69 pages. 1970. DM 8,– / $ 2.20

Vol. 165: J. M. Cohen, Stable Homotopy. V, 194 pages. 1970. DM 16,– / $ 4.40

Vol. 166: A. J. Silberger, PGL_2 over the p-adics: its Representations, Spherical Functions, and Fourier Analysis. VII, 202 pages. 1970. DM 18,– / $ 5.00

Vol. 167: Lavrentiev, Romanov and Vasiliev, Multidimensional Inverse Problems for Differential Equations. V, 59 pages. 1970. DM 10,– / $ 2.80

Vol. 168: F. P. Peterson, The Steenrod Algebra and its Applications: A conference to Celebrate N. E. Steenrod's Sixtieth Birthday. VII, 317 pages. 1970. DM 22,– / $ 6.10

Vol. 169: M. Raynaud, Anneaux Locaux Henséliens. V, 129 pages. 1970. DM 12,– / $ 3.30

Vol. 170: Lectures in Modern Analysis and Applications III. Edited by C. T. Taam. VI, 213 pages. 1970. DM 18,– / $ 5.00.

Vol. 171: Set-Valued Mappings, Selections and Topological Properties of 2^X. Edited by W. M. Fleischman. X, 110 pages. 1970. DM 12,– / $ 3.30

Vol. 172: Y.-T. Siu and G. Trautmann, Gap-Sheaves and Extension of Coherent Analytic Subsheaves. V, 172 pages. 1971. DM 16,– / $ 4.40

Vol. 173: J. N. Mordeson and B. Vinograde, Structure of Arbitrary Purely Inseparable Extension Fields. IV, 138 pages. 1970. DM 14,– / $ 3.90.

Vol. 174: B. Iversen, Linear Determinants with Applications to the Picard Scheme of a Family of Algebraic Curves. VI, 69 pages. 1970. DM 8,– / $ 2.20.

Vol. 175: M. Brelot, On Topologies and Boundaries in Potential Theory. VI, 176 pages. 1971. DM 18,– / $ 5.00

Vol. 176: H. Popp, Fundamentalgruppen algebraischer Mannigfaltigkeiten. IV, 154 Seiten. 1970. DM 16,– / $ 4.40

Vol. 177: J. Lambek, Torsion Theories, Additive Semantics and Rings of Quotients. VI, 94 pages. 1971. DM 12,– / $ 3.30

Vol. 178: Th. Bröcker und T. tom Dieck, Kobordismentheorie. XVI, 191 Seiten. 1970. DM 18,– / $ 5.00

Vol. 179: Seminaire Bourbaki – vol. 1968/69. Exposés 347-363. IV. 295 pages. 1971. DM 22,– / $ 6.10

Vol. 180: Séminaire Bourbaki – vol. 1969/70. Exposés 364-381. IV, 310 pages. 1971. DM 22,– / $ 6.10

Vol. 181: F. DeMeyer and E. Ingraham, Separable Algebras over Commutative Rings. V, 157 pages. 1971. DM 16,– / $ 4.40

Vol. 182: L. D. Baumert. Cyclic Difference Sets. VI, 166 pages. 1971. DM 16,– / $ 4.40

Vol. 183: Analytic Theory of Differential Equations. Edited by P. F. Hsieh and A. W. J. Stoddart. VI, 225 pages. 1971. DM 20,– / $ 5.50

Vol. 184: Symposium on Several Complex Variables, Park City, Utah, 1970. Edited by R. M. Brooks. V, 234 pages. 1971. DM 20,– / $ 5.50

Vol. 185: Several Complex Variables II, Maryland 1970. Edited by J. Horváth. III, 287 pages. 1971. DM 24,– / $ 6.60

Vol. 186: Recent Trends in Graph Theory. Edited by M. Capobianco/ J. B. Frechen/M. Krolik. VI, 219 pages. 1971. DM 18,– / $ 5.00

Vol. 187: H. S. Shapiro, Topics in Approximation Theory. VIII, 275 pages. 1971. DM 22,– / $ 6.10

Vol. 188: Symposium on Semantics of Algorithmic Languages. Edited by E. Engeler. VI, 372 pages. 1971. DM 26,– / $ 7.20

Vol. 189: A. Weil, Dirichlet Series and Automorphic Forms. V, 164 pages. 1971. DM 16,– / $ 4.40

Vol. 190: Martingales. A Report on a Meeting at Oberwolfach, May 17-23, 1970. Edited by H. Dinges. V, 75 pages. 1971. DM 12,– / $ 3.30

Vol. 191: Séminaire de Probabilités V. Edited by P. A. Meyer. IV, 372 pages. 1971. DM 26,– / $ 7.20

Vol. 192: Proceedings of Liverpool Singularities – Symposium I. Edited by C. T. C. Wall. V, 319 pages. 1971. DM 24,– / $ 6.60

Vol. 193: Symposium on the Theory of Numerical Analysis. Edited by J. Ll. Morris. VI, 152 pages. 1971. DM 16,– / $ 4.40

Vol. 194: M. Berger, P. Gauduchon et E. Mazet. Le Spectre d'une Variété Riemannienne. VII, 251 pages. 1971. DM 22,– / $ 6.10

Vol. 195: Reports of the Midwest Category Seminar V. Edited by J.W. Gray and S. Mac Lane.III, 255 pages. 1971. DM 22,– / $ 6.10

Vol. 196: H-spaces – Neuchâtel (Suisse)- Août 1970. Edited by F. Sigrist, V, 156 pages. 1971. DM 16,– / $ 4.40

Lecture Notes in Physics

Vol. 1: J. C. Erdmann, Wärmeleitung in Kristallen, theoretische Grundlagen und fortgeschrittene experimentelle Methoden. II, 283 Seiten. 1969. DM 20,– / $ 5.50

Vol. 2: K. Hepp, Théorie de la renormalisation. III, 215 pages. 1969. DM 18,– / $ 5.00

Vol. 3: A. Martin, Scattering Theory: Unitarity, Analytic and Crossing. IV, 125 pages. 1969. DM 14,– / $ 3.90

Vol. 4: G. Ludwig, Deutung des Begriffs physikalische Theorie und axiomatische Grundlegung der Hilbertraumstruktur der Quantenmechanik durch Hauptsätze des Messens. XI, 469 Seiten.1970. DM 28,– / $ 7.70

Vol. 5: M. Schaaf, The Reduction of the Product of Two Irreducible Unitary Representations of the Proper Orthochronous Quantummechanical Poincaré Group. IV, 120 pages. 1970. DM 14,– / $ 3.90

Vol. 6: Group Representations in Mathematics and Physics. Edited by V. Bargmann. V, 340 pages. 1970. DM 24,– / $ 6.60